THE
UMBRAL
CALCULUS

THE UMBRAL CALCULUS

STEVEN ROMAN

EMERITUS PROFESSOR OF MATHEMATICS
IRVINE, CALIFORNIA

DOVER PUBLICATIONS, INC.
MINEOLA, NEW YORK

Bibliographical Note

This Dover edition, first published in 2019, is an unabridged republication of the work originally published as Volume III in the "Pure and Applied Mathematics" series by Academic Press, Inc., New York, in 1984.

Library of Congress Cataloging-in-Publication Data

Names: Roman, Steven, author.
Title: The umbral calculus / Steven Roman (Emeritus Professor of Mathematics, Irvine, California).
Description: 2019 Dover edition. | Mineola, New York : Dover Publications, Inc., 2019. | Originally published: New York: Academic Press, Inc., 1984; first published by Dover Publications in 2005 and reissued in 2019. | Includes bibliographical references and index.
Identifiers: LCCN 2018051970| ISBN 9780486834139 | ISBN 0486834131
Subjects: LCSH: Calculus. | Polynomials.
Classification: LCC QA303 .R685 2019 | DDC 515—dc23
LC record available at https://lccn.loc.gov/2018051970

Manufactured in the United States by LSC Communications
83413101   2019
www.doverpublications.com

To Donna and to my parents

CONTENTS

vii

PREFACE

This monograph is intended to be an elementary introduction to the modern umbral calculus. Since we have in mind the largest possible audience, the only prerequisite is an acquaintance with the basic notions of algebra, and perhaps a dose of applied mathematics (such as differential equations) to help put the theory in some mathematical perspective.

The title of this work really should have been *The Modern Classical Umbral Calculus*. Within the past few years many, indeed infinitely many, distinct umbral calculi have begun to be studied. Actually, the existence of distinct umbral calculi was recognized in a vague way as early as the 1930s but seems to have remained largely ignored until the past decade.

In any case, we shall occupy the vast majority of our time in studying one particular umbral calculus—the one that dates back to the 1850s and that has received the attention (both good and bad) of mathematicians up to the present time. For this, we use the term classical umbral calculus. Only in the last chapter do we glimpse the newer, much less well established, nonclassical umbral calculi.

The classical umbral calculus, as it was from 1850 to about 1970, consisted primarily of a symbolic technique for the manipulation of sequences, whose mathematical rigor left much to be desired. To drive this point home one need only look at Eric Temple Bell's unsuccessful attempt (in 1940) to convince the mathematical community to accept the umbral calculus as a legitimate mathematical tool. (Even now some are still trying to achieve Eric Temple Bell's original goal.) This old-style umbral calculus was, however, useful in deriving certain mathematical results; but unfortunately these results had to be verified by a different, more rigorous method.

In the 1970s Gian-Carlo Rota, a mathematician with a superlative talent for handling just this sort of situation, began to construct a completely rigorous foundation for the theory—one that was based on the relatively modern ideas of a linear functional, a linear operator, and an adjoint. In 1977, the author was fortunate enough to join in on this development.

It is this modern classical umbral calculus that is the subject of the present monograph.

Perusal of the table of contents will give the reader an idea of the organization of the book; but let us make a few remarks in this regard. A choice had to be made between the present organization of Chapters 2–4 and the alternative of integrating these chapters by applying each new aspect of the theory to a running list of examples. We feel that the alternative approach has a tendency to minimize the effect of the theory, making it difficult to see just what the umbral calculus can do in a specific instance. On the other hand, we recognize that it can be difficult to remain motivated in the face of a large dose of theory, untempered by any examples. For this reason, we have included at the end of Chapter 2 a very brief discussion of some of the more accessible examples. Chapter 4 contains a more complete discussion of these and other examples. Let us emphasize, however, that we do not intend this book to be a treatise on any particular polynomial sequence, nor do we make any claims concerning the originality of the formulas contained herein. While Chapter 2 contains the definition and general properties of the principal object of study — the Sheffer sequence — it is Chapter 3 that really goes to the heart of the modern umbral method. In Chapter 6 we touch on some of the nonclassical umbral calculi, but only enough to whet the appetite for, it is hoped, a sequel to this volume.

Before we begin, we should like to express our gratitude to Professor Gian-Carlo Rota. His help and encouragement have proved invaluable over the years.

CHAPTER 1

INTRODUCTION

1. A DEFINITION OF THE CLASSICAL UMBRAL CALCULUS

Sequences of polynomials play a fundamental role in applied mathematics. Such sequences can be described in various ways, for example,

(1)   by orthogonality conditions:

$$\int_a^b p_n(x)p_m(x)w(x)\,dx = \delta_{n,m},$$

where $w(x)$ is a weight function and $\delta_{n,m} = 0$ or 1 according as $n \neq m$ or $n = m$;

(2)   as solutions to differential equations: for instance, the Hermite polynomials $H_n(x)$ satisfy the second-order linear differential equation

$$y'' - 2xy' + 2ny = 0;$$

(3)   by generating functions: for instance, the Bernoulli polynomials $B_n^{(\alpha)}(x)$ are characterized by

$$\left(\frac{t}{e^t - 1}\right)^\alpha e^{xt} = \sum_{k=0}^\infty \frac{B_k^{(\alpha)}(x)}{k!}t^k;$$

(4)   by recurrence relations: as an example, the exponential polynomials $\phi_n(x)$ satisfy

$$\phi_{n+1}(x) = x(\phi_n(x) + \phi_n'(x));$$

(5)   by operational formulas: for example, the Laguerre polynomials $L_n^{(\alpha)}(x)$ satisfy

$$L_n^{(\alpha)}(x) = x^{-\alpha}e^x D^n e^{-x} x^{n+\alpha}$$

(some put $n!\,L_n^{(\alpha)}(x)$ on the left).

One of the simplest classes of polynomial sequences, yet still large enough to include many important instances, is the class of Sheffer sequences (also known, in a slightly different form, as sequences of Sheffer A-type zero or poweroids). This class may be defined in many ways, most commonly by a generating function and, as Sheffer himself did, by a differential recurrence relation. Although we shall not adopt either of these means of definition, let us point out now that a sequence $s_n(x)$ is a Sheffer sequence if and only if its generating function has the form

$$\sum_{k=0}^{\infty} \frac{s_k(x)}{k!} t^k = A(t)e^{xB(t)},$$

where

$$A(t) = A_0 + A_1 t + A_2 t^2 + \cdots \qquad (A_0 \neq 0)$$

and

$$B(t) = B_1 t + B_2 t^2 + \cdots \qquad (B_1 \neq 0).$$

The Sheffer class contains such important sequences as those formed by

   (1)   the Hermite polynomials, which play an important role in applied mathematics and physics (such as Brownian motion and the Schrödinger wave equation);

   (2)   the Laguerre polynomials, which also play a key role in applied mathematics and physics (they are involved in solutions to the wave equation of the hydrogen atom);

   (3)   the Bernoulli polynomials, which find applications, for example, in number theory (evaluation of the Hurwitz zeta function, a generalization of the famous Riemann zeta function);

   (4)   the Abel polynomials, which have a connection with geometric probability (the random placement of nonoverlapping arcs on a circle);

   (5)   the central factorial polynomials, which play a role in the interpolation of functions.

Now to the point at hand. The modern classical umbral calculus can be described as a systematic study of the class of Sheffer sequences, made by employing the simplest techniques of modern algebra.

More explicitly, if P is the algebra of polynomials in a single variable, the set P^* of all linear functionals on P is usually thought of as a vector space (under pointwise operations). However, it is well known that a linear functional on P can be identified with a formal power series. In fact, there is a one-to-one correspondence between linear functionals on P and formal power series in a single variable. For example, we may associate to each linear functional L the power series $\sum_{k=0}^{\infty} L(x^k)t^k/k!$. But the set of formal power series is usually given the structure of an *algebra* (under formal addition and multiplication). This algebra structure, the additive part of which "agrees" with the vector space structure on P^*, can be "transferred" to P^*. The algebra so obtained is called the *umbral algebra*, and the umbral calculus is the study of this algebra.

As a first step in this direction, since P^* now has the structure of an algebra, we may consider, for two linear functionals L and M, the geometric sequence $M, ML, ML^2, ML^3, \ldots$. Then under mild conditions on L and M, the equations

$$ML^k\big(s_n(x)\big) = n!\,\delta_{n,k}$$

for $n, k \geq 0$ uniquely determine a sequence $s_n(x)$ of polynomials which turns out to be of Sheffer type, and, conversely, for any sequence $s_n(x)$ of Sheffer type there are linear functionals L and M for which the above equations hold. Thus we may characterize the class of Sheffer sequences by means of the umbral algebra. The resulting interplay between the umbral algebra and the algebra of polynomials allows for the natural development of some powerful adjointness properties wherein lies the real strength of the theory.

The umbral calculus is, to be sure, formal mathematics. By this we mean that limiting processes, such as the convergence of infinite series, play no role. Formal mathematics, much of which comes under the headings of combinatorics, the calculus of finite differences, the theory of special functions, and formal solutions to differential equations, is, in the opinion of some, staging a comeback after many years of neglect. It is our hope that the present work will aid in this comeback.

2. PRELIMINARIES

Since formal power series play a predominant role in the umbral calculus, we should set down some basic facts concerning their use. The simple proofs either can be supplied by the reader or can be gleaned from other sources, such as the paper of Niven [1].

Let C be a field of characteristic zero. Let $\mathscr{F}$ be the set of all formal power series in the variable t over C. Thus an element of $\mathscr{F}$ has the form

$$f(t) = \sum_{k=0}^{\infty} a_k t^k \tag{1.2.1}$$

for a_k in C. Two formal power series are equal if and only if the coefficients of like powers of t are equal. It is well known that if addition and multiplication are defined formally,

$$\sum_{k=0}^{\infty} a_k t^k + \sum_{k=0}^{\infty} b_k t^k = \sum_{k=0}^{\infty} (a_k + b_k) t^k,$$

$$\left(\sum_{k=0}^{\infty} a_k t^k \right) \left(\sum_{k=0}^{\infty} b_k t^k \right) = \sum_{k=0}^{\infty} \left(\sum_{j=0}^{k} a_j b_{k-j} \right) t^k,$$

then $\mathscr{F}$ is an algebra (with no zero divisors).

The *order* $o(f(t))$ of a power series $f(t)$ is the smallest integer k for which the coefficient of t^k does not vanish. We take $o(f(t)) = +\infty$ if $f(t) = 0$. It is easy to see that

$$o(f(t)g(t)) = o(f(t)) + o(g(t)),$$

$$o(f(t) + g(t)) \geq \min\{o(f(t)), o(g(t))\}.$$

The series $f(t)$ has a multiplicative inverse, denoted by $f(t)^{-1}$ or $1/f(t)$, if and only if $o(f(t)) = 0$. We shall then say that $f(t)$ is *invertible*.

Let $f_k(t)$ be a sequence in $\mathscr{F}$ and suppose $o(f_k(t)) \to \infty$ as $k \to \infty$. Then for any series

$$g(t) = \sum_{k=0}^{\infty} b_k t^k$$

we may form the new series

$$\sum_{k=0}^{\infty} b_k f_k(t)$$

that is a well-defined formal power series in t. In particular, if $f_k(t) = f(t)^k$, with $o(f(t)) \geq 1$, then the composition of $g(t)$ with $f(t)$ is

$$g(f(t)) = \sum_{k=0}^{\infty} b_k f(t)^k.$$

It is clear that $o(g(f(t))) = o(g(t))o(f(t))$.

If $o(f(t)) = 1$, then the formal power series $f(t)$ has a compositional inverse $\bar{f}(t)$ satisfying $f(\bar{f}(t)) = \bar{f}(f(t)) = t$. A series $f(t)$ for which $o(f(t)) = 1$ will be called a *delta series*. The sequence $f(t)^k$ of powers of a delta series forms a *pseudobasis* for $\mathscr{F}$. That is, for any series $g(t)$ in $\mathscr{F}$ there exists a unique sequence of constants a_k for which

$$g(t) = \sum_{k=0}^{\infty} a_k f(t)^k.$$

If $f(t)$ has the form (1.2.1), the formal derivative of $f(t)$ with respect to t is

$$\partial_t f(t) = \sum_{k=1}^{\infty} k a_k t^{k-1}.$$

We shall also use the notation $f'(t)$ for $\partial_t f(t)$.

We use the notation

$$\delta_{n,k} = \begin{cases} 1 & \text{if} \quad n = k, \\ 0 & \text{if} \quad n \neq k \end{cases}$$

for the Kronecker delta function. There is a variety of notations for the lower (or falling) factorial $x(x - 1) \cdots (x - n + 1)$. We shall use the notations

$$(x)_n = x(x - 1) \cdots (x - n + 1)$$

and

$$x^{(n)} = x(x + 1) \cdots (x + n - 1),$$

which are the usual ones in combinatorics and are also used in the calculus of finite differences. (Another common notation in the calculus of finite differences is $x^{(n)} = x(x - 1) \cdots (x - n + 1)$, and in the theory of special functions one sees $(x)_n = x(x + 1) \cdots (x + n - 1)$.)

We abbreviate the expression Section k of Chapter n as Section $n.k$. References to the bibliography will be made simply by using the author's last name and the citation number, as in Riordan [1].

CHAPTER 2

SHEFFER SEQUENCES

1. THE UMBRAL ALGEBRA

Let P be the algebra of polynomials in the single variable x over the field C of characteristic zero. Let P^* be the vector space of all linear functionals on P. We use the notation

$$\langle L \mid p(x) \rangle,$$

borrowed from physics, to denote the action of a linear functional L on a polynomial $p(x)$, and we recall that the vector space operations on P^* are defined by

$$\langle L + M \mid p(x) \rangle = \langle L \mid p(x) \rangle + \langle M \mid p(x) \rangle$$

and

$$\langle cL \mid p(x) \rangle = c \langle L \mid p(x) \rangle$$

for any constant c in C. Since a linear functional is uniquely determined by its action on a basis, L is uniquely determined by the sequence of constants $\langle L \mid x^n \rangle$.

As in Chapter 1, we let $\mathscr{F}$ denote the algebra of formal power series in the variable t over the field C. The formal power series

$$f(t) = \sum_{k=0}^{\infty} \frac{a_k}{k!} t^k \tag{2.1.1}$$

defines a linear functional on P by setting

$$\langle f(t) \mid x^n \rangle = a_n \tag{2.1.2}$$

6

for all $n \geq 0$. In particular,

$$\langle t^k \mid x^n \rangle = n! \, \delta_{n,k}.$$

Actually, any linear functional L in P^* has the form (2.1.1). For if

$$f_L(t) = \sum_{k=0}^{\infty} \frac{\langle L \mid x^k \rangle}{k!} t^k, \qquad (2.1.3)$$

then from (2.1.2) we get

$$\langle f_L(t) \mid x^n \rangle = \langle L \mid x^n \rangle,$$

and so as linear functionals $L = f_L(t)$.

Theorem 2.1.1 The map $L \to f_L(t)$ is a vector space isomorphism from P^* onto $\mathscr{F}$.

Proof Since $L = M$ if and only if $\langle L \mid x^k \rangle = \langle M \mid x^k \rangle$ for all $k \geq 0$, which holds if and only if $f_L(t) = f_M(t)$, the map is bijective. But we also have

$$f_{L+M}(t) = \sum_{k=0}^{\infty} \frac{\langle L + M \mid x^k \rangle}{k!} t^k$$

$$= \sum_{k=0}^{\infty} \frac{\langle L \mid x^k \rangle}{k!} t^k + \sum_{k=0}^{\infty} \frac{\langle M \mid x^k \rangle}{k!} t^k$$

$$= f_L(t) + f_M(t),$$

and, similarly,

$$f_{cL}(t) = c f_L(t).$$

Thus our map is a vector space isomorphism.

We shall obscure the isomorphism of Theorem 2.1.1 by thinking of linear functionals on P as formal power series in t, using (2.1.2) and (2.1.3) as our guiding light. Henceforth, $\mathscr{F}$ will denote both the algebra of formal power series in t and the vector space of all linear functionals on P, and so an element $f(t)$ of $\mathscr{F}$ will be thought of as both a formal power series and a linear functional. It is important to keep in mind that two elements $f(t)$ and $g(t)$ of $\mathscr{F}$ are equal as formal power series if and only if they are equal as linear functionals. This follows directly from (2.1.2) and the corresponding definitions of equality.

Thus we have automatically defined an algebra structure on the vector space of all linear functionals on P, namely, the algebra of formal power series. We shall call $\mathscr{F}$ the *umbral algebra*.

Let us give an example. For y in C the *evaluation functional* is defined to be the power series e^{yt}. From (2.1.2) we have

$$\langle e^{yt} \mid x^n \rangle = y^n,$$

and so

$$\langle e^{yt} \,|\, p(x) \rangle = p(y) \qquad (2.1.4)$$

for all $p(x)$ in P, which explains the name. If e^{zt} is evaluation at z, then

$$e^{yt} e^{zt} = e^{(y+z)t},$$

and so the product of evaluation at y with evaluation at z is evaluation at $y + z$.

Notice that for all $f(t)$ in $\mathscr{F}$

$$f(t) = \sum_{k=0}^{\infty} \frac{\langle f(t) \,|\, x^k \rangle}{k!} t^k \qquad (2.1.5)$$

and for all polynomials $p(x)$

$$p(x) = \sum_{k \geq 0} \frac{\langle t^k \,|\, p(x) \rangle}{k!} x^k. \qquad (2.1.6)$$

Let us consider some simple consequences of the results so far.

Proposition 2.1.2   If $f(t)$ and $g(t)$ are in $\mathscr{F}$, then

$$\langle f(t)g(t) \,|\, x^n \rangle = \sum_{k=0}^{n} \binom{n}{k} \langle f(t) \,|\, x^k \rangle \langle g(t) \,|\, x^{n-k} \rangle.$$

Proof   If we write both $f(t)$ and $g(t)$ in the form of (2.1.5), then the formal product is

$$f(t)g(t) = \sum_{m=0}^{\infty} \left(\sum_{k=0}^{m} \binom{m}{k} \langle f(t) \,|\, x^k \rangle \langle g(t) \,|\, x^{m-k} \rangle \right) \frac{t^m}{m!}.$$

By applying both sides of this, as linear functionals, to x^n and using (2.1.2), we get the desired result. ∎

It is easy to extend this result (by induction) to the product of several linear functionals.

Proposition 2.1.3   If $f_1(t), \ldots, f_m(t)$ are in $\mathscr{F}$, then

$$\langle f_1(t) \cdots f_m(t) \,|\, x^n \rangle = \sum \binom{n}{i_1, \ldots, i_m} \langle f_1(t) \,|\, x^{i_1} \rangle \cdots \langle f_m(t) \,|\, x^{i_m} \rangle,$$

where the sum is over all $i_1, \ldots, i_m$ such that $i_1 + \cdots + i_m = n$.

Proposition 2.1.4   If $o(f(t)) > \deg p(x)$, then $\langle f(t) \,|\, p(x) \rangle = 0$.

Proof   This follows easily from (2.1.2) since $\langle f(t) \,|\, x^n \rangle = 0$ whenever $o(f(t)) > n$. ∎

Proposition 2.1.5 If $o(f_k(t)) = k$ for all $k \geq 0$, then

$$\left\langle \sum_{k=0}^{\infty} a_k f_k(t) \,\bigg|\, p(x) \right\rangle = \sum_{k=0}^{\infty} a_k \langle f_k(t) \,|\, p(x) \rangle$$

for all $p(x)$ in P, the second sum being a finite one.

Proof Suppose that $\deg p(x) = d$. Then

$$\left\langle \sum_{k=0}^{\infty} a_k f_k(t) \,\bigg|\, p(x) \right\rangle = \left\langle \sum_{k=0}^{d} a_k f_k(t) + \sum_{k=d+1}^{\infty} a_k f_k(t) \,\bigg|\, p(x) \right\rangle$$

$$= \left\langle \sum_{k=0}^{d} a_k f_k(t) \,\bigg|\, p(x) \right\rangle$$

$$= \sum_{k=0}^{d} a_k \langle f_k(t) \,|\, p(x) \rangle$$

$$= \sum_{k=0}^{\infty} a_k \langle f_k(t) \,|\, p(x) \rangle.$$

Proposition 2.1.6 If $o(f_k(t)) = k$ for all $k \geq 0$ and if

$$\langle f_k(t) \,|\, p(x) \rangle = \langle f_k(t) \,|\, q(x) \rangle$$

for all k, then $p(x) = q(x)$.

Proof Since the sequence $f_k(t)$ forms a pseudobasis for $\mathscr{F}$, for all $n \geq 0$ there exist constants $a_{n,k}$ for which $t^n = \sum_{k=0}^{\infty} a_{n,k} f_k(t)$. Thus

$$\langle t^n \,|\, p(x) \rangle = \sum_{k=0}^{\infty} a_{n,k} \langle f_k(t) \,|\, p(x) \rangle$$

$$= \sum_{k=0}^{\infty} a_{n,k} \langle f_k(t) \,|\, q(x) \rangle$$

$$= \langle t^n \,|\, q(x) \rangle$$

and so (2.1.6) shows that $p(x) = q(x)$.

Proposition 2.1.6 implies that if $o(f_k(t)) = k$ and $\langle f_k(t) \,|\, p(x) \rangle = 0$ for all $k \geq 0$, then $p(x) = 0$.

Proposition 2.1.7 If $\deg p_k(x) = k$ for all $k \geq 0$ and if

$$\langle f(t) \,|\, p_k(x) \rangle = \langle g(t) \,|\, p_k(x) \rangle$$

for all k, then $f(t) = g(t)$.

Proof For each $n \geq 0$ there exist constants $a_{n,k}$ for which

$$x^n = \sum_{k=0}^{n} a_{n,k} p_k(x).$$

Thus

$$\langle f(t) \,|\, x^n \rangle = \sum_{k=0}^{n} a_{n,k} \langle f(t) \,|\, p_k(x) \rangle$$

$$= \sum_{k=0}^{n} a_{n,k} \langle g(t) \,|\, p_k(x) \rangle$$

$$= \langle g(t) \,|\, x^n \rangle,$$

and so (2.1.5) shows that $f(t) = g(t)$.

Proposition 2.1.7 implies that if $\deg p_k(x) = k$ and $\langle f(t) \,|\, p_k(x) \rangle = 0$ for all $k \geq 0$, then $f(t) = 0$.

The reader may have noticed that for any polynomial $p(x)$

$$\langle t^k \,|\, p(x) \rangle = p^{(k)}(0).$$

In words, the linear functional t^k is the kth derivative evaluated at 0. The multiplicative identity t^0 is simply evaluation at 0.

When we are considering a delta series $f(t)$ in $\mathscr{F}$ as a linear functional we will refer to it as a *delta functional*. Similarly, when we are considering an invertible series as a linear functional, we use the term *invertible functional*.

Proposition 2.1.8   The series $f(t)$ is a delta functional if and only if

$$\langle f(t)|1 \rangle = 0 \qquad \text{and} \qquad \langle f(t)|x \rangle \neq 0.$$

Proof   This follows directly from the definition of delta series and (2.1.5).

Proposition 2.1.9   The series $f(t)$ is an invertible functional if and only if $\langle f(t)|1 \rangle \neq 0$.

Let us now give our first umbral result.

Theorem 2.1.10   If $f(t)$ is in $\mathscr{F}$, then

$$\langle f(t) \,|\, xp(x) \rangle = \langle \partial_t f(t) \,|\, p(x) \rangle$$

for all polynomials $p(x)$.

Proof   By linearity we need only verify this for $p(x) = x^n$. But if

$$f(t) = \sum_{k=0}^{\infty} \frac{a_k}{k!} t^k,$$

then

$$\langle \partial_t f(t) \,|\, x^n \rangle = \left\langle \sum_{k=1}^{\infty} \frac{a_k}{(k-1)!} t^{k-1} \,\middle|\, x^n \right\rangle = a_{n+1}$$

$$= \left\langle \sum_{k=0}^{\infty} \frac{a_k}{k!} t^k \,\middle|\, x^{n+1} \right\rangle = \langle f(t) \,|\, x \cdot x^n \rangle.$$

The following simple result will be used frequently.

Proposition 2.1.11   For any $f(t)$ in $\mathscr{F}$ and $p(x)$ in P,

$$\langle f(t) \,|\, p(ax) \rangle = \langle f(at) \,|\, p(x) \rangle$$

for all constants a in C.

Proof   Taking $f(t) = t^n$ and $p(x) = x^k$, we have

$$\langle t^n \,|\, (ax)^k \rangle = a^k n! \, \delta_{n,k} = a^n n! \, \delta_{n,k} = \langle (at)^n \,|\, x^k \rangle.$$

This may be extended by linearity to all $f(t)$ and $p(x)$.

We conclude this section with a few examples of linear functionals that will appear in the sequel.

Example 1.   The *evaluation functional* is the invertible functional e^{yt} and we have

$$\langle e^{yt} \,|\, p(x) \rangle = p(y).$$

Example 2.   The *forward difference functional* is the delta functional $e^{yt} - 1$ and

$$\langle e^{yt} - 1 \,|\, p(x) \rangle = p(y) - p(0).$$

Example 3.   The *Abel functional* is the delta functional te^{yt}. We have

$$\langle te^{yt} \,|\, x^n \rangle = \left\langle \sum_{k=0}^{\infty} \frac{y^k}{k!} t^{k+1} \,\middle|\, x^n \right\rangle = n y^{n-1},$$

and so

$$\langle te^{yt} \,|\, p(x) \rangle = p'(y).$$

Example 4.   The invertible functional $(1 - t)^{-1}$ satisfies

$$\langle (1 - t)^{-1} \,|\, x^n \rangle = \left\langle \sum_{k=0}^{\infty} t^k \,\middle|\, x^n \right\rangle = n!.$$

It is associated with the Laguerre polynomials. Since

$$n! = \int_0^{\infty} u^n e^{-u} \, du,$$

we get the integral representation

$$\langle (1 - t)^{-1} \,|\, p(x) \rangle = \int_0^{\infty} p(u) e^{-u} \, du.$$

Example 5.   The functional $f(t)$ that satisfies

$$\langle f(t) \,|\, p(x) \rangle = \int_0^y p(u) \, du$$

for all polynomials $p(x)$ can be determined from (2.1.5) to be

$$f(t) = \sum_{k=0}^{\infty} \frac{\langle f(t) \mid x^k \rangle}{k!} t^k = \sum_{k=0}^{\infty} \frac{y^{k+1}}{(k+1)!} t^k = \frac{e^{yt} - 1}{t}.$$

Its inverse $t/(e^{yt} - 1)$ is associated with the Bernoulli polynomials. In fact, the numbers

$$B_n = \langle t/(e^t - 1) \mid x^n \rangle$$

are known as the Bernoulli numbers. We shall have more to say about this in Chapter 4.

Example 6. The invertible functional $(1 + e^{yt})/2$ satisfies

$$\langle (1 + e^{yt})/2 \mid p(x) \rangle = (p(0) + p(y))/2.$$

Its inverse $2/(1 + e^{yt})$ is associated with the Euler polynomials, also discussed in Chapter 4.

2. LINEAR OPERATORS

Let us begin this section by doing something that may seem, at first, to lead to some confusion. Namely, we use the notation t^k for the kth derivative operator on P, that is,

$$t^k x^n = \begin{cases} (n)_k x^{n-k}, & k \le n, \\ 0, & k > n, \end{cases}$$

where $(n)_k = n(n-1) \cdots (n-k+1)$. With this notation, any power series

$$f(t) = \sum_{k=0}^{\infty} \frac{a_k}{k!} t^k \tag{2.2.1}$$

is a linear operator on P defined by

$$f(t)x^n = \sum_{k=0}^{n} \binom{n}{k} a_k x^{n-k}. \tag{2.2.2}$$

Notice that we use juxtaposition $f(t)p(x)$ to denote the action of the operator $f(t)$ on the polynomial $p(x)$.

Thus an element of $\mathscr{F}$ plays three roles in the umbral calculus—it is a formal power series, a linear functional, and a linear operator. A little familiarity should remove any discomfort that may be felt by the use of this trinity, and the notational difference between

$$\langle f(t) \mid p(x) \rangle \qquad \text{and} \qquad f(t)p(x)$$

will make the particular role of $f(t)$ clear.

As with linear functionals, we note that two elements $f(t)$ and $g(t)$ in $\mathscr{F}$ are equal as formal power series if and only if they are equal as linear operators. (To see the "if" part, take successively $n = 0, 1, 2, \ldots$ in (2.2.2).)

Since $(t^k t^j)x^n = t^{k+j}x^n = t^k(t^j x^n)$, we conclude that

$$[f(t)g(t)]p(x) = f(t)[g(t)p(x)] \tag{2.2.3}$$

for all $f(t)$ and $g(t)$ in $\mathscr{F}$ and $p(x)$ in P, and so we may write $f(t)g(t)p(x)$ without ambiguity. Equation (2.2.3) shows that the product in $\mathscr{F}$ is also the composition of operators. We remark that

$$f(t)g(t)p(x) = g(t)f(t)p(x)$$

for all $f(t)$ and $g(t)$ in $\mathscr{F}$ and $p(x)$ in P.

The operator t^0 is, of course, the identity operator.

When we think of a delta (or invertible) series as an operator, we shall refer to it as a *delta* (or *invertible*) *operator*.

Let us consider the operator analogs of Propositions 2.1.4–2.1.7. Since the proofs are also analogous, we shall omit them.

Proposition 2.2.1   If $o(f(t)) > \deg p(x)$, then $f(t)p(x) = 0$.

Proposition 2.2.2   If $o(f_k(t)) = k$ for all $k \geq 0$, then

$$\left[\sum_{k=0}^{\infty} a_k f_k(t) \right] p(x) = \sum_{k=0}^{\infty} a_k [f_k(t)p(x)]$$

for all $p(x)$ in P, the second sum being a finite one.

Proposition 2.2.3   If $o(f_k(t)) = k$ for all $k \geq 0$ and if $f_k(t)p(x) = f_k(t)q(x)$ for all k, then $p(x) = q(x)$.

Proposition 2.2.4   If $\deg p_k(x) = k$ for all $k \geq 0$ and if $f(t)p_k(x) = g(t)p_k(x)$ for all k, then $f(t) = g(t)$.

We now come to the key relationship between the functional $f(t)$ and the operator $f(t)$.

Theorem 2.2.5   If $f(t)$ and $g(t)$ are in $\mathscr{F}$, then

$$\langle f(t)g(t) \mid p(x) \rangle = \langle g(t) \mid f(t)p(x) \rangle$$

for all polynomials $p(x)$.

Proof   Writing $f(t)$ and $g(t)$ in the form

$$f(t) = \sum_{k=0}^{\infty} \frac{\langle f(t) \mid x^k \rangle}{k!} t^k,$$

$$g(t) = \sum_{k=0}^{\infty} \frac{\langle g(t) \mid x^k \rangle}{k!} t^k$$

and using (2.2.2), we get

$$\langle g(t)\,|\,f(t)x^n\rangle = \left\langle g(t)\,\middle|\,\sum_{k=0}^{n}\binom{n}{k}\langle f(t)\,|\,x^k\rangle x^{n-k}\right\rangle$$

$$= \sum_{k=0}^{n}\binom{n}{k}\langle f(t)\,|\,x^k\rangle\langle g(t)\,|\,x^{n-k}\rangle,$$

which, according to Proposition 2.1.2, gives the desired result.

Incidentally, consideration of Theorems 2.1.10 and 2.2.5 points out the advantages of the notation $\langle L\,|\,p(x)\rangle$.

From Theorem 2.2.5 we see that

$$\langle f(t)\,|\,p(x)\rangle = \langle t^0\,|\,f(t)p(x)\rangle.$$

In words, applying the functional $f(t)$ to $p(x)$ is the same as applying the operator $f(t)$ and then evaluating at $x = 0$.

Let us consider the operator versions of the examples of Section 1.

1. The operator e^{yt} satisfies

$$e^{yt}x^n = \sum_{k=0}^{\infty}\frac{y^k}{k!}t^k x^n = \sum_{k=0}^{n}\binom{n}{k}y^k x^{n-k} = (x + y)^n,$$

and so

$$e^{yt}p(x) = p(x + y).$$

For this reason e^{yt} is called a *translation operator*.

2. The *forward difference operator* $e^{yt} - 1$ satisfies

$$(e^{yt} - 1)p(x) = p(x + y) - p(x).$$

3. The *Abel operator* is the delta operator te^{yt}, and we have

$$te^{yt}p(x) = tp(x + y) = p'(x + y).$$

4. The operator $(1 - t)^{-1}$ satisfies

$$(1 - t)^{-1}x^n = \sum_{k=0}^{\infty}(n)_k x^{n-k}.$$

Since

$$\int_0^{\infty}(x + u)^n e^{-u}\,du = \int_0^{\infty}\left[\sum_{k=0}^{n}\binom{n}{k}x^{n-k}u^k\right]e^{-u}\,du$$

$$= \sum_{k=0}^{n}k!\binom{n}{k}x^{n-k} = \sum_{k=0}^{n}(n)_k x^{n-k}$$

$$= \sum_{k=0}^{n}t^k x^n = (1 - t)^{-1}x^n,$$

we have

$$(1 - t)^{-1}p(x) = \int_0^\infty p(x + u)e^{-u}\,du.$$

5. The operator $(e^{yt} - 1)/t$ is easily seen to satisfy

$$\frac{e^{yt} - 1}{t}x^n = \int_x^{x+y} u^n\,du,$$

and so

$$\frac{e^{yt} - 1}{t}p(x) = \int_x^{x+y} p(u)\,du.$$

6. The operator $(1 + e^{yt})/2$ is given by

$$\frac{1 + e^{yt}}{2}p(x) = \frac{p(x) + p(x + y)}{2}.$$

Unlike the case for linear functionals, not all linear operators on P take the form of a power series in $\mathscr{F}$. For example, since $\deg f(t)p(x) \le \deg p(x)$ for any $f(t)$ in $\mathscr{F}$, the linear operator multiplication by x,

$$p(x) \to xp(x),$$

cannot have the form $f(t)$ for any series in $\mathscr{F}$. We devote the remainder of this section to various characterizations of the operators in $\mathscr{F}$. As a point of semantics, we shall say that a linear operator T on P has the form $g(t)$ if $Tp(x) = g(t)p(x)$ for all polynomials $p(x)$.

First we require a lemma.

Lemma 2.2.6 Let $f(t)$ be a delta operator and let T be a linear operator on P that commutes with $f(t)$, that is,

$$f(t)[Tp(x)] = T[f(t)p(x)]$$

for all polynomials $p(x)$. Then $\deg Tp(x) \le \deg p(x)$ for all $p(x)$.

Proof If the conclusion did not hold, there would exist an integer $m \ge 0$ such that $\deg Tx^m \ge m + 1$. Now, since $o(f(t)) = 1$, it is clear that

$$\deg f(t)p(x) = \deg p(x) - 1$$

whenever $\deg p(x) > 0$. Thus we have

$$0 = Tf(t)^{m+1}x^m = f(t)^{m+1}Tx^m \ne 0.$$

This contradiction establishes the lemma.

Theorem 2.2.7 A linear operator T on P has the form $g(t)$ if and only if it commutes with the derivative operator, that is, if and only if

$$T[tp(x)] = t[Tp(x)]$$

for all polynomials $p(x)$.

Proof   If T has the form $g(t)$, then it commutes with all operators in $\mathscr{F}$. For the converse, let

$$g(t) = \sum_{k=0}^{\infty} \frac{\langle t^0 \mid Tx^k \rangle}{k!} t^k.$$

Then

$$g(t)x^n = \sum_{k=0}^{\infty} \frac{\langle t^0 \mid Tx^k \rangle}{k!} t^k x^n = \sum_{k=0}^{n} \binom{n}{k} \langle t^0 \mid Tx^k \rangle x^{n-k}.$$

But, using Theorem 2.2.5, we see that

$$\langle t^0 \mid Tx^k \rangle = \frac{k!}{n!} \langle t^0 \mid Tt^{n-k}x^n \rangle$$

$$= \frac{k!}{n!} \langle t^0 \mid t^{n-k}Tx^n \rangle$$

$$= \frac{k!}{n!} \langle t^{n-k} \mid Tx^n \rangle,$$

and so

$$g(t)x^n = \sum_{k=0}^{n} \frac{\langle t^{n-k} \mid Tx^n \rangle}{(n-k)!} x^{n-k}$$

$$= \sum_{k=0}^{n} \frac{\langle t^k \mid Tx^n \rangle}{k!} x^k = Tx^n.$$

Thus T has the form $g(t)$ and the proof is complete.

Corollary 2.2.8   A linear operator T on P has the form $g(t)$ if and only if it commutes with any delta operator.

Proof   If T has the form $g(t)$, then it commutes with any operator in $\mathscr{F}$. Conversely, suppose that $Tf(t) = f(t)T$ for some delta operator $f(t)$. Then there exist constants a_k such that

$$t = \sum_{k=0}^{\infty} a_k f(t)^k.$$

Using Lemma 2.2.6, we have

$$Ttx^n = T \sum_{k=0}^{n} a_k f(t)^k x^n$$

$$= \sum_{k=0}^{n} a_k f(t)^k Tx^n = tTx^n$$

and so we may invoke Theorem 2.2.7 to conclude the proof.

Corollary 2.2.9　A linear operator T on P is of the form $g(t)$ if and only if it commutes with any translation operator e^{yt}.

Proof　This follows from Corollary 2.2.8 and the fact that T commutes with e^{yt} if and only if it commutes with the delta operator $e^{yt} - 1$.

3. SHEFFER SEQUENCES

By a sequence $s_n(x)$ of polynomials we shall always imply that $\deg s_n(x) = n$.

Theorem 2.3.1　Let $f(t)$ be a delta series and let $g(t)$ be an invertible series. Then there exists a unique sequence $s_n(x)$ of polynomials satisfying the orthogonality conditions

$$\langle g(t)f(t)^k \,|\, s_n(x)\rangle = n!\,\delta_{n,k} \qquad (2.3.1)$$

for all $n, k \geq 0$.

Proof　The uniqueness follows from Proposition 2.1.6. For the existence, if we set $s_n(x) = \sum_{j=0}^{n} a_{n,j} x^j$ and $g(t)f(t)^k = \sum_{i=k}^{\infty} b_{k,i} t^i$ with $b_{k,k} \neq 0$, then (2.3.1) becomes

$$n!\,\delta_{n,k} = \left\langle \sum_{i=k}^{\infty} b_{k,i} t^i \,\middle|\, \sum_{j=0}^{n} a_{n,j} x^j \right\rangle$$

$$= \sum_{i=k}^{\infty} \sum_{j=0}^{n} b_{k,i} a_{n,j} \langle t^i \,|\, x^j \rangle$$

$$= \sum_{i=k}^{n} b_{k,i} a_{n,i} i!.$$

Taking $k = n$, one obtains

$$a_{n,n} = 1/b_{n,n}.$$

By successively taking $k = n, n-1, \ldots, 0$, we obtain a triangular system of equations that can be solved for $a_{n,k}$.

We say that the sequence $s_n(x)$ in (2.3.1) is the *Sheffer sequence* for the pair $\big(g(t), f(t)\big)$, or that $s_n(x)$ is *Sheffer for* $\big(g(t), f(t)\big)$. Notice that $g(t)$ must be invertible and $f(t)$ must be a delta series.

There are two types of Sheffer sequences that deserve special consideration. The Sheffer sequence for $\big(1, f(t)\big)$ is the *associated sequence* for $f(t)$, and $s_n(x)$ is *associated to* $f(t)$. The Sheffer sequence for $\big(g(t), t\big)$ is the *Appell sequence* for $g(t)$, and $s_n(x)$ is *Appell for* $g(t)$. Incidentally, the term Appell sequence appears in the literature, but usually with a slightly different meaning; $s_n(x)$ is Appell in our sense if and only if $s_n(x)/n!$ is Appell according to these other sources. For convenience, in the next two sections we reformulate the results of this section for these two special cases.

Since $\langle t^k \,|\, x^n \rangle = n!\,\delta_{n,k}$, the sequence $p_n(x) = x^n$ is associated to the delta functional $f(t) = t$.

The next two results are among the most useful in the umbral calculus. The first one shows how to express any power series in terms of the geometric sequence $g(t)f(t)^k$.

Theorem 2.3.2   (The Expansion Theorem)   Let $s_n(x)$ be Sheffer for $\big(g(t), f(t)\big)$. Then for any $h(t)$ in $\mathscr{F}$

$$h(t) = \sum_{k=0}^{\infty} \frac{\langle h(t) \,|\, s_k(x) \rangle}{k!} g(t)f(t)^k.$$

Proof   The expansion theorem follows from Proposition 2.1.7 since

$$\left\langle \sum_{k=0}^{\infty} \frac{\langle h(t) \,|\, s_k(x) \rangle}{k!} g(t)f(t)^k \,\bigg|\, s_n(x) \right\rangle = \sum_{k=0}^{\infty} \frac{\langle h(t) \,|\, s_k(x) \rangle}{k!} \langle g(t)f(t)^k \,|\, s_n(x) \rangle$$

$$= \langle h(t) \,|\, s_n(x) \rangle.$$

The polynomial analog of the expansion theorem shows how to express an arbitrary polynomial as a linear combination of polynomials from a Sheffer sequence.

Theorem 2.3.3   (The Polynomial Expansion Theorem)   Let $s_n(x)$ be Sheffer for $\big(g(t), f(t)\big)$. Then for any polynomial $p(x)$ we have

$$p(x) = \sum_{k \geq 0} \frac{\langle g(t)f(t)^k \,|\, p(x) \rangle}{k!} s_k(x).$$

Proof   Applying $g(t)f(t)$ to either side of this equation gives the same result $\langle g(t)f(t)^k \,|\, p(x) \rangle$. Hence Proposition 2.1.6 completes the proof.

It is our intention now to characterize Sheffer sequences in several ways. We begin with the generating function.

Theorem 2.3.4   The sequence $s_n(x)$ is Sheffer for $\big(g(t), f(t)\big)$ if and only if

$$\frac{1}{g(\bar{f}(t))} e^{y\bar{f}(t)} = \sum_{k=0}^{\infty} \frac{s_k(y)}{k!} t^k \tag{2.3.2}$$

for all y in C, where $\bar{f}(t)$ is the compositional inverse of $f(t)$.

Proof   If $s_n(x)$ is Sheffer for $\big(g(t), f(t)\big)$, then by the expansion theorem

$$e^{yt} = \sum_{k=0}^{\infty} \frac{\langle e^{yt} \,|\, s_k(x) \rangle}{k!} g(t)f(t)^k = \sum_{k=0}^{\infty} \frac{s_k(y)}{k!} g(t)f(t)^k.$$

Thus

$$\frac{1}{g(t)}e^{yt} = \sum_{k=0}^{\infty}\frac{s_k(y)}{k!}f(t)^k$$

and finally

$$\frac{1}{g(\bar{f}(t))}e^{y\bar{f}(t)} = \sum_{k=0}^{\infty}\frac{s_k(y)}{k!}t^k.$$

For the converse, suppose that (2.3.2) holds. Then if $r_n(x)$ is Sheffer for $(g(t), f(t))$, we have

$$\sum_{k=0}^{\infty}\frac{r_k(y)}{k!}t^k = \frac{1}{g(\bar{f}(t))}e^{y\bar{f}(t)} = \sum_{k=0}^{\infty}\frac{s_k(y)}{k!}t^k,$$

and so $r_k(y) = s_k(y)$ for all y in C, which implies that $r_k(x) = s_k(x)$.

If we allow the use of formal power series in the two variables x and t, we could write (2.3.2) as

$$\frac{1}{g(\bar{f}(t))}e^{x\bar{f}(t)} = \sum_{k=0}^{\infty}\frac{s_k(x)}{k!}t^k. \tag{2.3.3}$$

While this is the usual form for a generating function, it has a small drawback in the present context, namely, when we think of x and t as formal variables they commute, but when we think of t as a linear operator they do not, for t acts on polynomials in x. This is not a serious problem and when we use (2.3.3) it is with the tacit understanding that t is nothing but a formal variable.

Some caution must be exercised with the term Sheffer sequence when consulting the literature. For example, sequences of A-type zero are characterized by Sheffer as sequences $u_n(x)$ that have a generating function of the form

$$A(t)e^{xB(t)} = \sum_{k=0}^{\infty}u_k(x)t^k,$$

where $o(A(t)) = 0$ and $o(B(t)) = 1$. Thus $s_n(x)$ is a Sheffer sequence if and only if $s_n(x)/n!$ is a sequence of Sheffer A-type zero. As we have already noted, the same remark applies to Appell sequences.

Incidentally, sequences of Sheffer A-type zero are called poweroids by Steffensen [1] and sequences of generalized Appell type by the Bateman Manuscript Project (Erdelyi [1]), although in Boas and Buck [1] the latter term is used for a more general set of polynomial sequences.

The generating function leads us to a representation for Sheffer sequences.

Theorem 2.3.5 The sequence $s_n(x)$ is Sheffer for $(g(t), f(t))$ if and only if

$$s_n(x) = \sum_{k=0}^{n}\frac{1}{k!}\langle g(\bar{f}(t))^{-1}f(t)^k \mid x^n\rangle x^k. \tag{2.3.4}$$

Proof   Applying the right side of (2.3.2) to x^n gives

$$\left\langle \sum_{k=0}^{\infty} \frac{s_k(y)}{k!} t^k \,\middle|\, x^n \right\rangle = s_n(y),$$

and applying the left side to x^n gives

$$\langle g(\bar{f}(t))^{-1} e^{y\bar{f}(t)} \,|\, x^n \rangle = \left\langle \sum_{k=0}^{\infty} \frac{1}{k!} y^k g(\bar{f}(t))^{-1} f(t)^k \,\middle|\, x^n \right\rangle$$

$$= \sum_{k=0}^{n} \frac{1}{k!} \langle g(\bar{f}(t))^{-1} \bar{f}(t)^k \,|\, x^n \rangle y^k.$$

Since this holds for all y in C, the result follows.

Equation (2.3.4) is called the *conjugate representation* for $s_n(x)$.

Theorem 2.3.6   The sequence $s_n(x)$ is Sheffer for $(g(t), f(t))$ if and only if $g(t)s_n(x)$ is the associated sequence for $f(t)$.

Proof   This follows directly from Theorem 2.2.5 and the definitions since

$$\langle f(t)^k \,|\, g(t)s_n(x) \rangle = \langle g(t)f(t)^k \,|\, s_n(x) \rangle = n! \, \delta_{n,k}.$$

Theorem 2.3.6 says that each associated sequence gives rise to a class of Sheffer sequences, one sequence for each invertible operator $g(t)$ in $\mathscr{F}$.

We next give an operator characterization of Sheffer sequences.

Theorem 2.3.7   A sequence $s_n(x)$ is Sheffer for $(g(t), f(t))$, for some invertible $g(t)$, if and only if

$$f(t)s_n(x) = ns_{n-1}(x) \tag{2.3.5}$$

for all $n \geq 0$.

Proof   If $s_n(x)$ is Sheffer for $(g(t), f(t))$, then

$$\langle g(t)f(t)^k \,|\, f(t)s_n(x) \rangle = \langle g(t)f(t)^{k+1} \,|\, s_n(x) \rangle$$

$$= n! \, \delta_{n,k+1}$$

$$= n(n-1)! \, \delta_{n-1,k}$$

$$= \langle g(t)f(t)^k \,|\, ns_{n-1}(x) \rangle,$$

and so $f(t)s_n(x) = ns_{n-1}(x)$. Conversely, suppose that (2.3.5) holds and let $p_n(x)$ be associated to $f(t)$. We define a linear operator T on P by

$$Ts_n(x) = p_n(x).$$

This operator is well defined and invertible since both $s_n(x)$ and $p_n(x)$ form a basis for P. Then, since the first part of this proof shows that

$f(t)p_n(x) = np_{n-1}(x)$, we have

$$Tf(t)s_n(x) = nTs_{n-1}(x)$$
$$= np_{n-1}(x)$$
$$= f(t)p_n(x)$$
$$= f(t)Ts_n(x),$$

and so $Tf(t) = f(t)T$. From Corollary 2.2.8 we deduce the existence of an invertible series $g(t)$ for which $g(t)s_n(x) = p_n(x)$. The result follows from Theorem 2.3.6.

From Theorems 2.3.6 and 2.3.7 we get a formula for the action of an operator in $\mathscr{F}$ on a Sheffer sequence.

Theorem 2.3.8 Let $s_n(x)$ be Sheffer for $(g(t), f(t))$ and let $p_n(x)$ be associated to $f(t)$. Then for any $h(t)$ in $\mathscr{F}$,

$$h(t)s_n(x) = \sum_{k=0}^{n} \binom{n}{k} \langle h(t) \mid s_k(x) \rangle p_{n-k}(x).$$

Proof By the expansion theorem

$$h(t) = \sum_{k=0}^{\infty} \frac{\langle h(t) \mid s_k(x) \rangle}{k!} g(t)f(t)^k.$$

Applying each side of this equation, as an operator, to $s_n(x)$ gives

$$h(t)s_n(x) = \sum_{k=0}^{\infty} \frac{\langle h(t) \mid s_k(x) \rangle}{k!} g(t)f(t)^k s_n(x)$$

$$= \sum_{k=0}^{n} \binom{n}{k} \langle h(t) \mid s_k(x) \rangle g(t)s_{n-k}(x)$$

$$= \sum_{k=0}^{n} \binom{n}{k} \langle h(t) \mid s_k(x) \rangle p_{n-k}(x).$$

We turn now to an algebraic characterization of Sheffer sequences that generalizes the binomial formula.

Theorem 2.3.9 (The Sheffer Identity) A sequence $s_n(x)$ is Sheffer for $(g(t), f(t))$, for some invertible $g(t)$, if and only if

$$s_n(x + y) = \sum_{k=0}^{n} \binom{n}{k} p_k(y)s_{n-k}(x)$$

for all y in C, where $p_n(x)$ is associated to $f(t)$.

Proof Suppose that $s_n(x)$ is Sheffer for $(g(t), f(t))$. By the expansion theorem

$$e^{yt} = \sum_{k=0}^{\infty} \frac{p_k(y)}{k!} f(t)^k,$$

and applying both sides of this equation to $s_n(x)$ gives

$$s_n(x + y) = e^{yt} s_n(x)$$

$$= \sum_{k=0}^{\infty} \frac{p_k(y)}{k!} f(t)^k s_n(x)$$

$$= \sum_{k=0}^{\infty} \binom{n}{k} p_k(y) s_{n-k}(x).$$

For the converse, suppose that the sequence $s_n(x)$ satisfies the Sheffer identity, where $p_n(x)$ is associated to $f(t)$. We define a linear operator T on P by

$$T s_n(x) = p_n(x).$$

Then according to Theorem 2.3.6 it suffices to prove that T has the form $g(t)$ in $\mathscr{F}$. The first part of this proof shows that $p_n(x)$ satisfies the Sheffer identity and so

$$e^{yt} T s_n(x) = e^{yt} p_n(x)$$

$$= p_n(x + y)$$

$$= \sum_{k=0}^{n} \binom{n}{k} p_k(y) p_{n-k}(x)$$

$$= T \sum_{k=0}^{n} \binom{n}{k} p_k(y) s_{n-k}(x)$$

$$= T s_n(x + y)$$

$$= T e^{yt} s_n(x).$$

Therefore $e^{yt} T = T e^{yt}$ and Corollary 2.2.9 shows that T has the form $g(t)$ in $\mathscr{F}$. This concludes the proof.

Incidentally, by interchanging x and y in the Sheffer identity and setting $y = 0$, we get

$$s_n(x) = \sum_{k=0}^{n} \binom{n}{k} s_{n-k}(0) p_k(x).$$

Thus, given a sequence $p_n(x)$ associated to $f(t)$, each Sheffer sequence $s_n(x)$ that uses $f(t)$ as its delta functional is uniquely determined by the sequence of constant terms $s_n(0)$. Moreover, any sequence of constants a_n for which $a_0 \neq 0$

gives rise in this way to a Sheffer sequence, using $f(t)$ as its delta functional. To see this, suppose that a_n is such a sequence of constants and define $s_n(x)$ by

$$s_n(x) = \sum_{k=0}^{n} \binom{n}{k} a_{n-k} p_k(x).$$

Then since $a_0 \neq 0$, $s_n(x)$ is a sequence and

$$\begin{aligned}
f(t)s_n(x) &= f(t) \sum_{k=0}^{n} \binom{n}{k} a_{n-k} p_k(x) \\
&= \sum_{k=1}^{n} \binom{n}{k} k a_{n-k} p_{k-1}(x) \\
&= n \sum_{k=1}^{n} \binom{n-1}{k-1} a_{n-1-(k-1)} p_{k-1}(x) \\
&= n \sum_{k=0}^{n-1} \binom{n-1}{k} a_{n-1-k} p_k(x) \\
&= n s_{n-1}(x).
\end{aligned}$$

By Theorem 2.3.7 the sequence $s_n(x)$ is Sheffer for $\big(g(t), f(t)\big)$ for some $g(t)$.

An important property of Sheffer sequences is their performance with respect to multiplication in $\mathscr{F}$. One might wish to compare the next result with Proposition 2.1.2.

Theorem 2.3.10 Let $s_n(x)$ be Sheffer for $\big(g(t), f(t)\big)$ and let $p_n(x)$ be associated to $f(t)$. Then for any $h(t)$ and $l(t)$ in $\mathscr{F}$ we have

$$\langle h(t)l(t) \mid s_n(x) \rangle = \sum_{k=0}^{n} \binom{n}{k} \langle h(t) \mid s_k(x) \rangle \langle l(t) \mid p_{n-k}(x) \rangle.$$

Proof According to Theorem 2.3.8

$$\begin{aligned}
\langle h(t)l(t) \mid s_n(x) \rangle &= \langle l(t) \mid h(t)s_n(x) \rangle \\
&= \left\langle l(t) \,\middle|\, \sum_{k=0}^{n} \binom{n}{k} \langle h(t) \mid s_k(x) \rangle p_{n-k}(x) \right\rangle \\
&= \sum_{k=0}^{n} \binom{n}{k} \langle h(t) \mid s_k(x) \rangle \langle l(t) \mid p_{n-k}(x) \rangle.
\end{aligned}$$

We conclude this section with an illustration of how nicely umbral results can marry to produce useful formulas that are general enough to apply to all Sheffer sequences.

We wish to determine the coefficients $a_{n,k}$ in the expansion

$$xs_n(x) = \sum_{k=0}^{n+1} a_{n,k} s_k(x),$$

where $s_n(x)$ is a Sheffer sequence. If $s_n(x)$ is Sheffer for $(g(t), f(t))$, then by the polynomial expansion theorem

$$xs_n(x) = \sum_{k=0}^{n+1} \frac{\langle g(t)f(t)^k \mid xs_n(x) \rangle}{k!} s_k(x).$$

By Theorems 2.1.10, 2.2.5, and 2.3.7 we deduce that

$$
\begin{aligned}
\langle g(t)f(t)^k \mid xs_n(x) \rangle &= \langle \partial_t g(t)f(t)^k \mid s_n(x) \rangle \\
&= \langle g'(t)f(t)^k + kg(t)f(t)^{k-1}f'(t) \mid s_n(x) \rangle \\
&= \langle g'(t) \mid f(t)^k s_n(x) \rangle \\
&\quad + k\langle g(t)f'(t) \mid f(t)^{k-1} s_n(x) \rangle \\
&= (n)_k \langle g'(t) \mid s_{n-k}(x) \rangle \\
&\quad + k(n)_{k-1} \langle g(t)f'(t) \mid s_{n-k+1}(x) \rangle.
\end{aligned}
$$

Thus we obtain the following expansion formula, which we shall use in Chapter 4.

Theorem 2.3.11   If $s_n(x)$ is Sheffer for $(g(t), f(t))$, then

$$xs_n(x) = \sum_{k=0}^{n+1} \left[\binom{n}{k} \langle g'(t) \mid s_{n-k}(x) \rangle + \binom{n}{k-1} \langle g(t)f'(t) \mid s_{n-k+1}(x) \rangle \right] s_k(x),$$

where we take $\binom{m}{j} = 0$ if $j < 0$ or $j > m$.

It is also worth singling out a formula for $s'_n(x)$ in terms of $s_k(x)$.

Theorem 2.3.12   If $s_n(x)$ is Sheffer for $(g(t), f(t))$ and $p_n(x)$ is associated to $f(t)$, then

$$s'_n(x) = \sum_{k=0}^{n-1} \binom{n}{k} \langle t \mid p_{n-k}(x) \rangle s_k(x).$$

Proof   By the polynomial expansion theorem

$$
\begin{aligned}
s'_n(x) &= \sum_{k=0}^{n-1} \frac{\langle g(t)f(t)^k \mid ts_n(x) \rangle}{k!} s_k(x) \\
&= \sum_{k=0}^{n-1} \frac{\langle t \mid g(t)f(t)^k s_n(x) \rangle}{k!} s_k(x) \\
&= \sum_{k=0}^{n-1} \binom{n}{k} \langle t \mid p_{n-k}(x) \rangle s_k(x).
\end{aligned}
$$

4. ASSOCIATED SEQUENCES

We reformulate the results of the previous section for associated sequences. One should take special note of Theorem 2.4.5.

Theorem 2.4.1 (The Expansion Theorem) Let $p_n(x)$ be associated to $f(t)$. Then for any $h(t)$ in $\mathscr{F}$

$$h(t) = \sum_{k=0}^{\infty} \frac{\langle h(t) \mid p_k(x) \rangle}{k!} f(t)^k.$$

Theorem 2.4.2 (The Polynomial Expansion Theorem) Let $p_n(x)$ be associated to $f(t)$. Then for any polynomial $p(x)$ we have

$$p(x) = \sum_{k \geq 0} \frac{\langle f(t)^k \mid p(x) \rangle}{k!} p_k(x).$$

Theorem 2.4.3 The sequence $p_n(x)$ is associated to $f(t)$ if and only if

$$e^{y\bar{f}(t)} = \sum_{k=0}^{\infty} \frac{p_k(y)}{k!} t^k$$

for all y in C.

Theorem 2.4.4 The sequence $p_n(x)$ is associated to $f(t)$ if and only if

$$p_n(x) = \sum_{k=1}^{n} \frac{\langle \bar{f}(t)^k \mid x^n \rangle}{k!} x^k.$$

This is the *conjugate representation* for associated sequences.

The operator characterization of associated sequences adds a new feature.

Theorem 2.4.5 The sequence $p_n(x)$ is associated to $f(t)$ if and only if

(i) $\langle t^0 \mid p_n(x) \rangle = \delta_{n,0}$,

(ii) $f(t)p_n(x) = np_{n-1}(x)$.

Proof Suppose that $p_n(x)$ is associated to $f(t)$. Then $\langle f(t)^k \mid p_n(x) \rangle = n! \, \delta_{n,k}$, and setting $k = 0$ gives (i). Part (ii) follows from Theorem 2.3.7. Conversely, if (i) and (ii) hold, then

$$\langle f(t)^k \mid p_n(x) \rangle = \langle t^0 \mid f(t)^k p_n(x) \rangle$$
$$= \langle t^0 \mid (n)_k p_{n-k}(x) \rangle$$
$$= (n)_k \delta_{n-k,0} = n! \, \delta_{n,k}.$$

One should notice that (i) is equivalent to $p_0(x) = 1$ and $p_n(0) = 0$ for $n > 0$.

Theorem 2.4.6   If $p_n(x)$ is associated to $f(t)$, then for any $h(t)$ in $\mathscr{F}$

$$h(t)p_n(x) = \sum_{k=0}^{n} \binom{n}{k} \langle h(t) \,|\, p_k(x) \rangle p_{n-k}(x).$$

Theorem 2.4.7   (The Binomial Identity)   The sequence $p_n(x)$ is an associated sequence if and only if

$$p_n(x + y) = \sum_{k=0}^{n} \binom{n}{k} p_k(y) p_{n-k}(x)$$

for all y in C.

A sequence $p_n(x)$ that satisfies the binomial identity is known as a sequence of *binomial type*. According to Theorem 2.4.7, a sequence is an associated sequence if and only if it is of binomial type.

Theorem 2.4.8   If $p_n(x)$ is associated to $f(t)$, then

$$xp_n(x) = \sum_{k=1}^{n+1} \binom{n}{k-1} \langle f'(t) \,|\, p_{n-k+1}(x) \rangle p_k(x).$$

Theorem 2.4.9   If $p_n(x)$ is an associated sequence, then

$$p_n'(x) = \sum_{k=0}^{n-1} \binom{n}{k} \langle t \,|\, p_{n-k}(x) \rangle p_k(x).$$

5. APPELL SEQUENCES

Let us recast the results of Section 3 for Appell sequences. Recall that our Appell sequences may differ from others in the literature by a factor of $n!$. The last result of this section has not appeared previously.

Theorem 2.5.1   (The Expansion Theorem)   Let $s_n(x)$ be the Appell sequence for $g(t)$. Then for any $h(t)$ in $\mathscr{F}$

$$h(t) = \sum_{k=0}^{\infty} \frac{\langle h(t) \,|\, s_k(x) \rangle}{k!} g(t) t^k.$$

Theorem 2.5.2   (The Polynomial Expansion Theorem)   Let $s_n(x)$ be Appell for $g(t)$. Then for any polynomial $p(x)$ we have

$$p(x) = \sum_{k \geq 0} \frac{\langle g(t) \,|\, p^{(k)}(x) \rangle}{k!} s_k(x),$$

where $p^{(k)}(x) = t^k p(x)$ is the kth derivative of $p(x)$.

Theorem 2.5.3   The sequence $s_n(x)$ is Appell for $g(t)$ if and only if

$$\frac{1}{g(t)}e^{yt} = \sum_{k=0}^{\infty} \frac{s_k(y)}{k!} t^k$$

for all y in C.

Theorem 2.5.4   The sequence $s_n(x)$ is Appell for $g(t)$ if and only if

$$s_n(x) = \sum_{k=0}^{n} \binom{n}{k} \langle g(t)^{-1} \mid x^{n-k} \rangle x^k.$$

This is the *conjugate representation* for Appell sequences.

Theorem 2.5.5   The sequence $s_n(x)$ is Appell for $g(t)$ if and only if

$$s_n(x) = g(t)^{-1} x^n.$$

Theorem 2.5.6   The sequence $s_n(x)$ is Appell for $g(t)$ if and only if

$$ts_n(x) = ns_{n-1}(x),$$

that is,

$$s_n'(x) = ns_{n-1}(x).$$

It is common in the literature to find that Appell sequences are defined by the condition $u_n'(x) = u_{n-1}(x)$. Then $n!\, u_n(x)$ is Appell in our sense.

Theorem 2.5.7   Let $s_n(x)$ be Appell for $g(t)$. Then for any $h(t)$ in $\mathscr{F}$

$$h(t)\, s_n(x) = \sum_{k=0}^{n} \binom{n}{k} \langle h(t) \mid s_k(x) \rangle x^{n-k}.$$

Theorem 2.5.8   (The Appell Identity)   The sequence $s_n(x)$ is an Appell sequence if and only if

$$s_n(x + y) = \sum_{k=0}^{n} \binom{n}{k} s_k(y) x^{n-k}.$$

Theorem 2.5.9   If $s_n(x)$ is Appell for $g(t)$, then

$$xs_n(x) = s_{n+1}(x) + \sum_{k=0}^{n} \binom{n}{k} \langle g'(t) \mid s_{n-k}(x) \rangle s_k(x).$$

Theorem 2.5.10   (The Multiplication Theorem)   If $s_n(x)$ is the Appell sequence for $g(t)$, then for any constant $\alpha \neq 0$

$$s_n(\alpha x) = \alpha^n \frac{g(t)}{g(t/\alpha)} s_n(x)$$

for all $n \geq 0$.

Proof　First we recall that, according to Proposition 2.1.11, for any $f(t)$ and $p(x)$

$$\langle f(t) \mid p(x) \rangle = \langle f(t/\alpha) \mid p(\alpha x) \rangle.$$

Then we have

$$\langle t^k \mid g(t/\alpha) s_n(\alpha x) \rangle = \langle \alpha^k t^k g(t) \mid s_n(x) \rangle$$
$$= \alpha^k n! \, \delta_{n,k}$$
$$= \alpha^n n! \, \delta_{n,k}$$
$$= \langle t^k \mid \alpha^n g(t) s_n(x) \rangle,$$

and so

$$g(t/\alpha) s_n(\alpha x) = \alpha^n g(t) s_n(x),$$

from which the result follows.

6. A FEW EXAMPLES

For reasons outlined in the Preface, we pause briefly to discuss a few examples of Sheffer sequences. No proofs or derivations will be given here and for convenience all results will be repeated in Chapter 4.

Although it may seem from the coming discussion that in some cases the determination of the Sheffer sequence for a given pair of linear functionals is oblique, in the next chapter we shall obtain formulas for the explicit computation of these sequences.

One may wish to refer to the examples of linear functionals at the end of Section 2.1 and the corresponding examples of operators in Section 2.2.

Example 1.　The sequence $p_n(x) = x^n$ is, of course, associated to the delta functional $f(t) = t$. The generating function

$$\sum_{k=0}^{\infty} \frac{x^k}{k!} t^k = e^{xt}$$

and the binomial identity

$$(x + y)^n = \sum_{k=0}^{n} \binom{n}{k} x^k y^{n-k}$$

should come as no surprise.

Example 2.　The lower factorial polynomials

$$(x)_n = x(x - 1) \cdots (x - n + 1),$$

also called the falling factorial or binomial polynomials, are associated to the forward difference functional

$$f(t) = e^t - 1.$$

This can easily be seen from Theorem 2.4.5.

The generating function is

$$\sum_{k=0}^{\infty} \frac{(x)_k}{k!} t^k = e^{x \log(1+t)},$$

which is equivalent to the binomial expansion

$$\sum_{k=0}^{\infty} \binom{x}{k} t^k = (1+t)^x.$$

The binomial identity is

$$(x+y)_n = \sum_{k=0}^{n} \binom{n}{k} (x)_k (y)_{n-k},$$

which can be rewritten as

$$\binom{x+y}{n} = \sum_{k=0}^{n} \binom{x}{k} \binom{y}{n-k}.$$

This is known as Vandermonde's convolution formula.

Example 3. The Abel polynomials

$$A_n(x;a) = x(x-an)^{n-1}$$

are associated to the Abel functional

$$f(t) = te^{at}.$$

This can readily be verified from Theorem 2.4.5.

The generating function is

$$\sum_{k=0}^{\infty} \frac{x(x-ak)^{k-1}}{k!} t^k = e^{x\bar{f}(t)}.$$

Incidentally, if we differentiate with respect to x, we obtain

$$\sum_{k=0}^{\infty} \frac{k(x-a)(x-ak)^{k-1}}{k!} t^k = \bar{f}(t) e^{x\bar{f}(t)}.$$

Setting $x = 0$ gives

$$\sum_{k=1}^{\infty} \frac{(-a)^k k^{k-1}}{(k-1)!} t^k = \bar{f}(t).$$

The binomial identity in this setting is

$$(x + y)(x + y - an)^{n-1} = \sum_{k=0}^{n} \binom{n}{k} xy(x - ak)^{k-1}(y - a(n - k))^{n-k-1},$$

which is a formula due to Abel.

Example 4.   The Hermite polynomials $H_n(x)$ form the Appell sequence for

$$g(t) = e^{t^2/2}.$$

By Theorem 3.5.5 we have

$$H_n(x) = e^{-t^2/2}x^n = \sum_{k=0}^{\infty} \left(-\frac{1}{2}\right)^k \frac{1}{k!} t^{2k} x^n$$

$$= \sum_{k \geq 0} \left(-\frac{1}{2}\right)^k \frac{(n)_{2k}}{k!} x^{n-k}.$$

The generating function is

$$\sum_{k=0}^{\infty} \frac{H_k(x)}{k!} t^k = e^{-t^2/2} e^{xt} = e^{xt - (t^2/2)}.$$

The Appell identity (Theorem 2.5.8) is

$$H_n(x + y) = \sum_{k=0}^{n} \binom{n}{k} H_k(x) y^{n-k}$$

Some care must be exercised here since the term Hermite polynomial is frequently used for the sequence $s_n(x)$ satisfying

$$\sum_{k=0}^{\infty} \frac{s_k(x)}{k!} = e^{2xt - t^2}.$$

This sequence is not exactly Appell, but it is Sheffer for $(e^{t^2/4}, t/2)$.

In Chapter 4 we discuss the more general Hermite polynomials $H_n^{(v)}(x)$ of variance v.

Example 5.   The Bernoulli polynomials $B_n(x)$ form the Appell sequence for

$$g(t) = (e^t - 1)/t.$$

Theorem 2.5.5 gives us

$$B_n(x) = \left(t/(e^t - 1)\right)x^n.$$

The generating function is

$$\sum_{k=0}^{\infty} \frac{B_k(x)}{k!} t^k = \frac{t}{e^t - 1} e^{xt}.$$

Taking $x = 0$, we have

$$\sum_{k=0}^{\infty} \frac{B_k(0)}{k!} t^k = \frac{t}{e^t - 1}.$$

The numbers $B_n(0)$ are known as the Bernoulli numbers.

In Chapter 4 we discuss the more general Bernoulli polynomials $B_n^{(\alpha)}(x)$ of order α.

Example 6. The Laguerre polynomials $L_n^{(\alpha)}(x)$ of order α (sometimes known as the generalized Laguerre polynomials) form the Sheffer sequence for

$$g(t) = (1 - t)^{-\alpha - 1},$$

$$f(t) = t/(t - 1).$$

The sequence $L_n^{(-1)}$ is associated to $f(t) = t/(t - 1)$.

From the formulas of the next chapter we shall easily compute that

$$L_n^{(\alpha)}(x) = \sum_{k=0}^{n} \frac{n!}{k!} \binom{\alpha + n}{n - k}(-x)^k.$$

The generating function is

$$\sum_{k=0}^{\infty} \frac{L_k^{(\alpha)}(x)}{k!} t^k = \frac{1}{(1 - t)^{\alpha + 1}} e^{xt/(t - 1)}.$$

The Sheffer identity is

$$L_n^{(\alpha)}(x + y) = \sum_{k=0}^{n} \binom{n}{k} L_k^{(\alpha)}(x) L_{n-k}^{(-1)}(y).$$

Many interesting properties of the Laguerre polynomials follow by umbral methods from the fact that $\bar{f}(t) = f(t)$.

Once again, some care must be exercised with the term Laguerre polynomials, for in the literature they are frequently defined as $L_n^{(\alpha)}(x)/n!$. Also, for reasons related to orthogonality, the condition $\alpha > -1$ is sometimes invoked. Needless to say, we shall disregard this constraint.

CHAPTER 3

OPERATORS AND THEIR ADJOINTS

1. CONTINUOUS OPERATORS ON P^*

In this and the next two sections, we shall let P^* denote the umbral algebra, which we have been denoting by $\mathscr{F}$. This will conform to standard usage and should not be misleading since, for the time being, we shall be concerned with the vector space structure of the umbral algebra only.

Linear operators defined on the umbral algebra P^* play an important role in the umbral calculus. An indispensible property of such an operator T is that it pass under infinite sums, that is,

$$T \sum_{k=0}^{\infty} a_k f_k(t) = \sum_{k=0}^{\infty} a_k T f_k(t) \tag{3.1.1}$$

whenever $\sum_{k=0}^{\infty} a_k f_k(t)$ exists, which, as we have remarked, is precisely when $o(f_k(t)) \to \infty$ as $k \to \infty$. But in order for (3.1.1) to be meaningful, the right side must also exist, that is, $o(T f_k(t)) \to \infty$ as $k \to \infty$. Thus a necessary condition for T to have the property expressed in (3.1.1) is that

$$o(T f_k(t)) \to \infty \qquad \text{whenever} \quad o(f_k(t)) \to \infty. \tag{3.1.2}$$

This turns out to be sufficient as well.

Theorem 3.1.1 If T is a linear operator on P^* satisfying (3.1.2), then

$$T \sum_{k=0}^{\infty} a_k f_k(t) = \sum_{k=0}^{\infty} a_k T f_k(t)$$

for all sequences $f_k(t)$ for which $o(f_k(t)) \to \infty$ as $k \to \infty$.

Proof Suppose that $o(f_k(t)) \to \infty$. For any polynomial $p(x)$ and any $m \geq 0$ we have

$$\left\langle T \sum_{k=0}^{\infty} a_k f_k(t) \,\middle|\, p(x) \right\rangle = \left\langle T \sum_{k=0}^{m} a_k f_k(t) \,\middle|\, p(x) \right\rangle + \left\langle T \sum_{k=m+1}^{\infty} a_k f_k(t) \,\middle|\, p(x) \right\rangle.$$

(3.1.3)

Now $o(f_k(t)) \to \infty$ implies that $o[\sum_{k=m+1}^{\infty} a_k f_k(t)] \to \infty$ as $m \to \infty$, and so by (3.1.2) we have $o[T \sum_{k=m+1}^{\infty} a_k f_k(t)] \to \infty$ as $m \to \infty$. Thus we may take m large enough so that the second term on the right of (3.1.3) vanishes and so that $o(Tf_k(t)) > \deg p(x)$ for $k > m$. Then

$$\left\langle T \sum_{k=0}^{\infty} a_k f_k(t) \,\middle|\, p(x) \right\rangle = \left\langle T \sum_{k=0}^{m} a_k f_k \,\middle|\, p(x) \right\rangle$$

$$= \left\langle \sum_{k=0}^{m} a_k Tf_k(t) \,\middle|\, p(x) \right\rangle$$

$$= \left\langle \sum_{k=0}^{\infty} a_k Tf_k(t) \,\middle|\, p(x) \right\rangle.$$

This proves the theorem.

We shall say that a linear operator T on P^* is *continuous* if it has the property expressed in (3.1.2).

Actually, it is possible to define a topology on P^* by specifying that the sequence $f_k(t)$ converges to $f(t)$ if, for any $n \geq 0$, there exists an integer k_n such that if $k > k_n$, then $f_k(t)$ and $f(t)$ agree in the first n terms; in symbols, $\langle f_k(t) | x^j \rangle = \langle f(t) | x^j \rangle$ for $0 \leq j \leq n - 1$. It can be shown that this makes P^* into a topological vector space. Clearly, the sequence $f_k(t)$ converges to 0 if and only if $o(f_k(t)) \to \infty$ as $k \to \infty$. Now a linear operator T is continuous if and only if $Tf_k(t)$ converges to 0 whenever $f_k(t)$ converges to 0. In other words, T is continuous if and only if it satisfies (3.1.2). Since we shall not require any topological notions other than continuity, we take (3.1.2) as the definition.

2. AUTOMORPHISMS AND DERIVATIONS ON P^*

Let us recall that a linear operator T on P^* is an algebra morphism if it preserves the product on P^*,

$$T[f(t)g(t)] = Tf(t)Tg(t).$$

If in addition T is bijective, then it is said to be an automorphism of P^*.

Theorem 3.2.1   If T is an automorphism of P^*, then T preserves order, that is,

$$o(Tf(t)) = o(f(t))$$

for all $f(t)$ in P^*.

Proof   First we observe that $Tf(t)$ is invertible if and only if $f(t)$ is invertible. For if $f(t)^{-1}$ exists, then $Tf(t)T(f(t)^{-1}) = T(f(t)f(t)^{-1} = T1 = 1$. Hence $Tf(t)$ is invertible, with inverse $T(f(t)^{-1})$. The converse follows in a similar manner, using the automorphism T^{-1}. In terms of order we have shown that $o(Tf(t)) = 0$ if and only if $o(f(t)) = 0$. In particular then, $o(Tt) \geq 1$. But if $o(Tt) > 1$, then there would be no $g(t)$ in P^* for which $Tg(t) = t$. For any $g(t)$ in P^* can be written in the form $g(t) = g_0 + tg_1(t)$, and then $Tg(t) = g_0 + T(tg_1(t)) \neq t$ because $o(Ttg_1(t)) = o(TtTg_1(t)) \geq o(Tt) > 1$. Thus, since T is surjective, we must have $o(Tt) = 1$. This implies that $o(Tt^k) = k$ for all $k \geq 0$. Finally, if $o(f(t)) = k$, then $f(t) = t^k g(t)$, where $o(g(t)) = 0$, and so $o(Tf(t)) = o(Tt^k Tg(t)) = o(Tt^k) + o(Tg(t)) = k$.

Corollary 3.2.2   Any automorphism of P^* is continuous.

As a consequence, we remark that any automorphism of P^* is uniquely determined by its value on any delta functional, in particular by its value on t.

A linear operator ∂ on P^* is a *derivation* on P^* if

$$\partial[f(t)g(t)] = f(t)\,\partial g(t) + g(t)\,\partial f(t)$$

for all $f(t)$ and $g(t)$ in P^*.

Theorem 3.2.3   If ∂ is a surjective derivation on P^*, then $\partial c = 0$ for all constants c in C and $o(\partial f(t)) = k - 1$ whenever $o(f(t)) = k \geq 1$.

Proof   First we observe that $\partial 1 = \partial 1^2 = \partial 1 + \partial 1$, and so $\partial 1 = 0$. Thus $\partial c = 0$ for any constant c. Now any $g(t)$ in P^* has the form $g(t) = g_0 + tg_1(t)$. Hence $\partial g(t) = t\,\partial g_1(t) + g_1(t)\,\partial t$, and since $o(t\,\partial g_1(t)) \geq 1$, the fact that ∂ is surjective implies $o(\partial t) = 0$. Finally, if $o(f(t)) = k \geq 1$, then $f(t) = t^k g(t)$, where $o(g(t)) = 0$, and so $o(\partial f(t)) = o(t^k\,\partial g(t) + kt^{k-1}g(t)\,\partial t) = k - 1$.

Corollary 3.2.4   Any surjective derivation of P^* is continuous.

3. ADJOINTS

If λ is a linear operator on P, its *adjoint* is the linear operator λ^* on P^* defined, for each $f(t)$ in P^*, by the condition

$$\langle \lambda^* f(t) \,|\, p(x) \rangle = \langle f(t) \,|\, \lambda p(x) \rangle \tag{3.3.1}$$

for all polynomials $p(x)$.

For example, since $\langle f(t)g(t) \,|\, p(x) \rangle = \langle g(t) \,|\, f(t)p(x) \rangle$, we see that the adjoint of the operator $\lambda = f(t)$ is the operator multiplication by $f(t)$. That is,

$$f(t)^* g(t) = f(t)g(t)$$

for all $g(t)$ in P^*.

It follows easily from (3.3.1) that

$$(\lambda + \mu)^* = \lambda^* + \mu^*,$$

$$(c\lambda)^* = c\lambda^*,$$

$$(\lambda \circ \mu)^* = \mu^* \circ \lambda^*,$$

$$(\lambda^{-1})^* = (\lambda^*)^{-1}, \quad \lambda \quad \text{invertible}.$$

(3.3.2)

Thus the adjoint map, which sends a linear operator λ on P to its adjoint λ^* on P^*, is a linear transformation from the vector space of all linear operators on P to the vector space of all linear operators on P^*. Furthermore, if $\lambda^* = 0$, then $\langle f(t) \mid \lambda p(x) \rangle = 0$ for all $f(t)$ in P^* and all $p(x)$ in P. Thus $\lambda = 0$ and the adjoint map is injective. In the next theorem we determine its range.

Theorem 3.3.1 A linear operator T on P^* is the adjoint of a linear operator λ on P if and only if T is continuous.

Proof Suppose first that $T = \lambda^*$ for some operator λ on P, and let

$$o(f_k(t)) \to \infty.$$

Then for all $n \geq 0$ there exists an integer k_n such that $k > k_n$ implies $o(f_k(t)) > \deg \lambda x^j$ for all $0 \leq j \leq n$. Thus if $k > k_n$, we have $\langle Tf_k(t) \mid x^j \rangle = \langle f_k(t) \mid \lambda x^j \rangle = 0$ for $0 \leq j \leq n$, that is, $o(Tf_k(t)) > n$. Thus $o(Tf_k(t)) \to \infty$ and T is continuous. Conversely, suppose that T is continuous. We define a linear operator λ on P by

$$\lambda x^n = \sum_{k \geq 0} \frac{\langle Tt^k \mid x^n \rangle}{k!} x^k,$$

the sum being a finite one since $o(Tt^k) > n$ for k large. Then

$$\langle \lambda^* t^m \mid x^n \rangle = \langle t^m \mid \lambda x^n \rangle$$

$$= \sum_{k \geq 0} \frac{\langle Tt^k \mid x^n \rangle}{k!} \langle t^m \mid x^k \rangle$$

$$= \langle Tt^m \mid x^n \rangle,$$

and so $\lambda^* t^m = Tt^m$ for all $m \geq 0$. Since both λ^* and T are continuous,

$$\lambda^* f(t) = Tf(t)$$

for all $f(t)$ in P^*. Thus $T = \lambda^*$ and the proof is complete.

In summary, we have established the following theorem.

Theorem 3.3.2 The adjoint map (Fig. 3.1), which sends a linear operator on P to its adjoint λ^* on P^*, is a vector space isomorphism from the vector space of

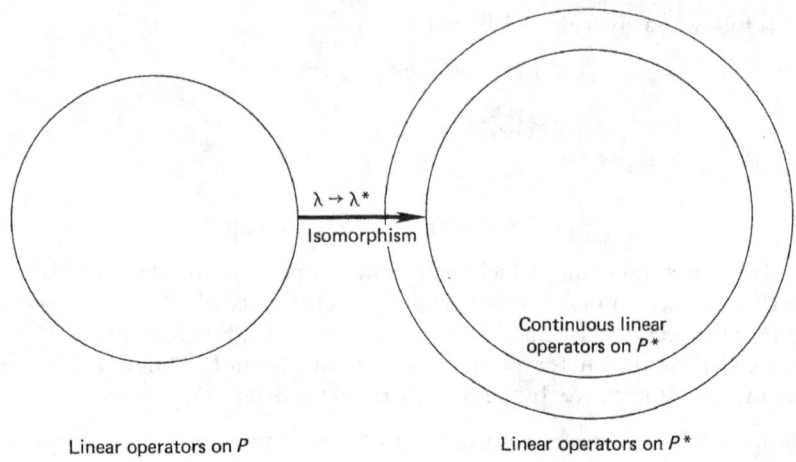

FIGURE 3.1   The adjoint map of Theorem 3.3.2.

all linear operators on *P* onto the vector space of all continuous linear operators on *P**.

Since both automorphisms and derivations on *P* are continuous, we may augment Fig. 3.1 as shown in Fig. 3.2.

The linear operators on *P* that fill in the upper left-hand box in Fig. 3.2 are known as umbral operators and those that fill in the lower left-hand box are known as umbral shifts. We shall study these important operators in Sections 4 and 6, but first let us attend to one more result.

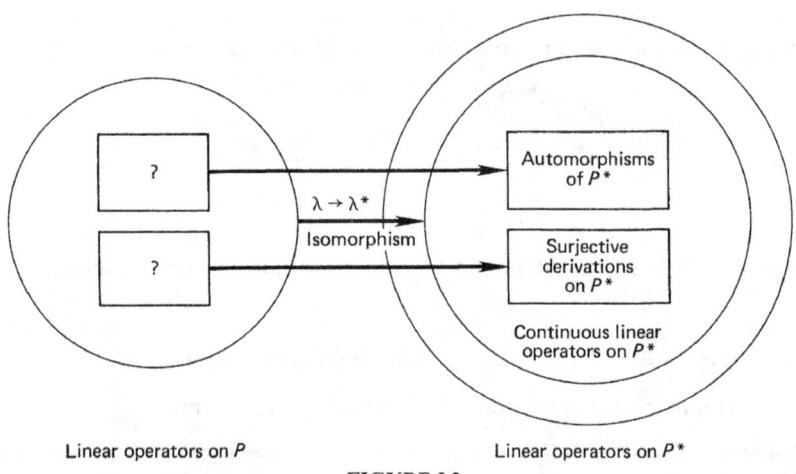

FIGURE 3.2

If T is a linear operator on P^*, then it too has an adjoint T^*, defined on P^{**} by

$$(T^*\phi)f(t) = \phi\big(Tf(t)\big)$$

for all ϕ in P^{**} and $f(t)$ in P^*.

Let us recall that while the vector spaces P and P^{**} are *not* isomorphic, there is a vector space isomorphism θ, from P *into* P^{**}, defined by

$$\big(\theta p(x)\big)f(t) = \langle f(t)\,|\,p(x)\rangle$$

for all $f(t)$ in P^* and $p(x)$ in P. The isomorphism θ is known as the canonical isomorphism. It is common to identify the image $\theta(P)$ of the canonical isomorphism with the space P of polynomials, in other words, to think of a polynomial $p(x)$ as a linear functional on P^*, where $p(x)f(t) = \langle f(t)\,|\,p(x)\rangle$.

If T is a linear operator on P^*, it is natural to wonder whether the adjoint $T^*: P^{**} \to P^{**}$ maps "polynomials" in P^{**} to "polynomials" in P^{**}; that is, whether $T^*\theta(P) \subset \theta(P)$. The answer is given in the next theorem. Notice that it is in terms of the original operator T.

Theorem 3.3.3 Let T be a linear operator on P^*. Then the adjoint T^* satisfies $T^*\theta(P) \subset \theta(P)$ if and only if T is continuous.

Proof Assume first that $T^*\theta(P) \subset \theta(P)$ and let $o\big(f_k(t)\big) \to \infty$. Then $\theta^{-1}T^*\theta x^j$ is in P for all $j \geq 0$, and for all $n \geq 0$ there exists an integer k_n such that $k > k_n$ implies $(T^*\theta x^j)f_k(t) = \langle f_k(t)\,|\,\theta^{-1}T^*\theta x^j\rangle = 0$ for $0 \leq j \leq n$. Thus $k > k_n$ implies $\langle Tf_k(t)\,|\,x^j\rangle = (\theta x^j)\big(Tf_k(t)\big) = (T^*\theta x^j)f_k(t) = 0$ for $0 \leq j \leq n$, that is, $o\big(Tf_k(t)\big) > n$. Hence T is continuous. For the converse, assume that T is continuous. Then $T = \lambda^*$ for some operator λ on P. We have

$$\big(T^*\theta p(x)\big)f(t)\big) = \big(\lambda^{**}\theta p(x)\big)f(t)$$
$$= \big(\theta p(x)\big)\big(\lambda^*f(t)\big)$$
$$= \langle \lambda^*f(t)\,|\,p(x)\rangle$$
$$= \langle f(t)\,|\,\lambda p(x)\rangle$$
$$= \big(\theta\lambda p(x)\big)f(t),$$

and so $T^*\theta p(x) = \theta\lambda p(x)$ and $T^*\theta = \theta\lambda$. Hence $T^*\theta(P) = \theta\lambda(P) \subset \theta(P)$ and the proof is complete.

4. UMBRAL OPERATORS AND UMBRAL COMPOSITION

We return to the notation $\mathscr{F}$ for the umbral algebra.

Let $p_n(x)$ be the associated sequence for $f(t)$. Then the linear operator λ on P, defined by

$$\lambda x^n = p_n(x),$$

is called the *umbral operator* for $p_n(x)$ or for $f(t)$. To indicate the dependence of λ on $f(t)$ we may use the notation λ_f.

Incidentally, in the next section we shall study *Sheffer operators*, which are linear operators $\lambda: x^n \to s_n(x)$, where $s_n(x)$ is an arbitrary Sheffer sequence.

Since $\deg p_n(x) = n$, it is clear that an umbral operator is a bijection.

It is our intention in this section to characterize umbral operators by means of their adjoints.

According to Theorem 3.3.1, the adjoint of an umbral operator is continuous. Much more is true, however, since, if $\lambda_f: x^n \to p_n(x)$ is an umbral operator, then

$$\langle \lambda_f^* f(t)^k \mid x^n \rangle = \langle f(t)^k \mid \lambda_f x^n \rangle$$
$$= \langle f(t)^k \mid p_n(x) \rangle$$
$$= n! \, \delta_{n,k}$$
$$= \langle t^k \mid x^n \rangle,$$

and so

$$\lambda_f^* f(t)^k = t^k$$

for all $k \geq 0$. The continuity of λ_f^* then implies that

$$\lambda_f^* g(f(t)) = g(t)$$

for all $g(t)$ in $\mathscr{F}$. In particular, if we take $g(t) = \bar{f}(t)^k$, then

$$\lambda_f^* t^k = \bar{f}(t)^k,$$

which leads to

$$\lambda_f^* g(t) = g(\bar{f}(t)) \tag{3.4.1}$$

for all $g(t)$ in $\mathscr{F}$. In words, the operator λ_f^* is substitution by $\bar{f}(t)$.

From (3.4.1) it follows that

$$\lambda_f^* g(t)h(t) = g(\bar{f}(t))h(\bar{f}(t)) = \lambda_f^* g(t)\lambda_f^* h(t),$$

and so λ_f^* is an automorphism of P. It is a pleasant fact that this characterizes umbral operators.

Theorem 3.4.1 A linear operator λ on P is an umbral operator if and only if its adjoint λ^* is an automorphism of $\mathscr{F}$. Moreover, if λ_f is an umbral operator, then

$$\lambda_f^* g(t) = g(\bar{f}(t))$$

for all $g(t)$ in $\mathscr{F}$. In particular, $\lambda_f^* f(t) = t$.

Proof We have already seen that the adjoint of an umbral operator is an automorphism of $\mathscr{F}$ satisfying (3.4.1). For the converse, suppose that λ^* is an automorphism of $\mathscr{F}$. It follows from Theorem 3.2.1 and the fact that λ^* is a bijection that there exists a unique delta series $f(t)$ for which $\lambda^* f(t) = t$. If $p_n(x)$ is associated to $f(t)$, then

$$\langle f(t)^k \,|\, \lambda x^n \rangle = \langle \lambda^* f(t)^k \,|\, x^n \rangle$$
$$= \langle t^k \,|\, x^n \rangle$$
$$= n!\, \delta_{n,k}$$
$$= \langle f(t)^k \,|\, p_n(x) \rangle,$$

and so $\lambda x^n = p_n(x)$. Therefore λ is an umbral operator and the proof is complete.

Corollary 3.2.2 and Theorems 3.3.1 and 3.4.1 imply that any automorphism of $\mathscr{F}$ is indeed the adjoint of an umbral operator. Thus we have the following result.

Theorem 3.4.2 The adjoint map, which sends a linear operator on P to its adjoint λ^* on $\mathscr{F}$, is a bijection between the set of all umbral operators on P and the set of all automorphisms of $\mathscr{F}$.

We can now fill in one of the boxes in Fig. 3.2 of Section 3 (see Fig. 3.3).

It is easy to see that the set of all automorphisms of $\mathscr{F}$ is a group under composition of operators. The adjoint map of Theorem 3.4.2 can be used to transfer this group structure to the set of umbral operators on P.

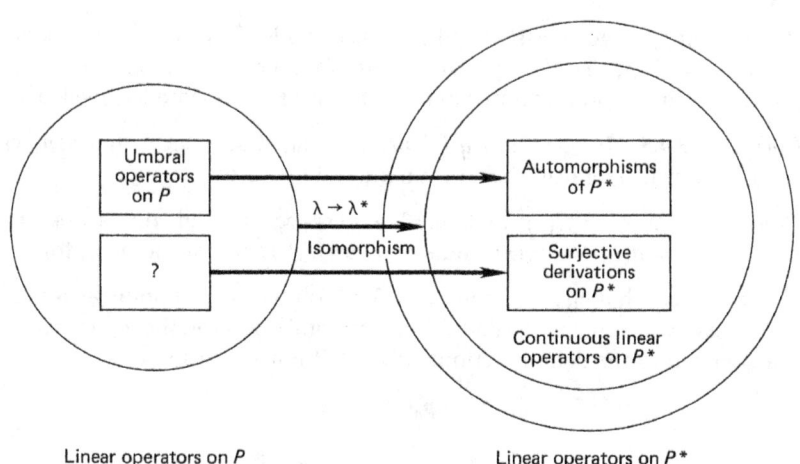

FIGURE 3.3

Theorem 3.4.3 The set of all umbral operators on P is a group under composition. In fact,

$$\lambda_f \circ \lambda_g = \lambda_{g \circ f}, \qquad \lambda_f^{-1} = \lambda_{\bar{f}}. \tag{3.4.2}$$

Proof We give two proofs of the first part of this theorem. The first one uses Theorem 3.4.2. Let λ and μ be umbral operators. Then $(\lambda \circ \mu)^* = \mu^* \circ \lambda^*$ is the composition of two automorphisms of $\mathscr{F}$ and so is an automorphism of $\mathscr{F}$. Thus by Theorem 3.4.2 the map $\lambda \circ \mu$ is an umbral operator. Similarly, since $(\lambda^{-1})^* = (\lambda^*)^{-1}$ is an automorphism of $\mathscr{F}$, the map λ^{-1} is also an umbral operator. Thus the umbral operators form a group under composition. The second proof is more direct and establishes (3.4.2) but is less elegant. We have from (3.4.1)

$$
\begin{aligned}
(\lambda_f \circ \lambda_g)^* h(t) &= \lambda_g^* \circ \lambda_f^* h(t) \\
&= \lambda_g^* h(\bar{f}(t)) \\
&= h(\bar{f}(\bar{g}(t))) \\
&= h((\overline{g \circ f})(t)) \\
&= \lambda_{g \circ f}^* h(t),
\end{aligned}
$$

so $(\lambda_f \circ \lambda_g)^* = \lambda_{g \circ f}^*$ and $\lambda_f \circ j_g = \lambda_{g \circ f}$. Similarly,

$$(\lambda_f^{-1})^* h(t) = (\lambda_f^*)^{-1} h(t) = h(f(t)) = \lambda_{\bar{f}} h(t),$$

and so $\lambda_f^{-1} = \lambda_{\bar{f}}$.

Corollary 3.4.4 An umbral operator maps associated sequences to associated sequences.

Proof Let $p_n(x)$ be an associated sequence and let λ be an umbral operator. Then if $\mu: x^n \to p_n(x)$ is the umbral operator for $p_n(x)$, we have $\lambda p_n(x) = \lambda \circ \mu x^n$. But $\lambda \circ \mu$ is an umbral operator and so $\lambda p_n(x)$ is an associated sequence.

Corollary 3.4.5 If $p_n(x)$ and $q_n(x)$ are associated sequences and $\alpha: p_n(x) \to q_n(x)$ is a linear operator, then α is an umbral operator.

Proof Let $\lambda: x^n \to p_n(x)$ and $\mu: x^n \to q_n(x)$ be umbral operators. Then $\mu \circ \lambda^{-1} p_n(x) = q_n(x) = \alpha p_n(x)$, and so $\alpha = \mu \circ \lambda^{-1}$ is an umbral operator.

Let us see what effect Theorem 3.4.3 has on the corresponding associated sequences. According to the definition of umbral operator, the sequence $\lambda_{g \circ f} x^n$ is associated to the delta functional $g(f(t))$. But if we write

$$\lambda_f : x^n \to p_n(x),$$

$$\lambda_g : x^n \to q_n(x) = \sum_{k=0}^{n} q_{n,k} x^k,$$

then

$$\lambda_{g \circ f} x^n = \lambda_f \circ \lambda_g x^n = \lambda_f q_n(x) = \sum_{k=0}^{n} q_{n,k} p_k(x). \tag{3.4.3}$$

In general, if $p_n(x)$ and $q_n(x) = \sum_{k=0}^{n} q_{n,k} x^k$ are sequences of polynomials, we define the *umbral composition of* $q_n(x)$ *with* $p_n(x)$ to be the sequence

$$q_n(\mathbf{p}(x)) = \sum_{k=0}^{n} q_{n,k} p_k(x).$$

Thus (3.4.3) may be written as

$$\lambda_{g \circ f} x^n = q_n(\mathbf{p}(x)). \tag{3.4.4}$$

Theorem 3.4.6 The set of associated sequences is a group under umbral composition. In particular, if $p_n(x)$ is associated to $f(t)$ and $q_n(x)$ is associated to $g(t)$, then $q_n(\mathbf{p}(x))$ is associated to $g(f(t))$. The identity under umbral composition is the sequence x^n and the inverse of the sequence $p_n(x)$ is the associated sequence for $\bar{f}(t)$.

Proof The second statement follows directly from (3.4.4). It is clear that the sequence x^n is the identity under umbral composition. Setting $g(t) = \bar{f}(t)$ in (3.4.4) gives

$$x^n = \lambda_{\bar{f} \circ f} x^n = q_n(\mathbf{p}(x)),$$

and so $\lambda_{\bar{f}} x^n = q_n(x)$ is the inverse of $p_n(x)$. But $\lambda_{\bar{f}} x^n$ is the associated sequence for $\bar{f}(t)$.

We close this section with a formula for the action of an automorphism of $\mathscr{F}$.

Theorem 3.4.7 Let λ be an umbral operator on P. Then as operators on P we have

$$\lambda^* g(t) = \lambda^{-1} g(t) \lambda$$

for any operator $g(t)$ in $\mathscr{F}$.

Proof For any $f(t)$ in $\mathscr{F}$ and $p(x)$ in P,

$$\begin{aligned}
\langle f(t) \,|\, (\lambda^* g(t)) p(x) \rangle &= \langle f(t)(\lambda^* g(t)) \,|\, p(x) \rangle \\
&= \langle \lambda^* [g(t)(\lambda^*)^{-1} f(t)] \,|\, p(x) \rangle \\
&= \langle g(t)(\lambda^*)^{-1} f(t) \,|\, \lambda p(x) \rangle \\
&= \langle (\lambda^{-1})^* f(t) \,|\, g(t) \lambda p(x) \rangle \\
&= \langle f(t) \,|\, \lambda^{-1} g(t) \lambda p(x) \rangle,
\end{aligned}$$

and so $\lambda^* g(t) p(x) = \lambda^{-1} g(t) \lambda p(x)$. Thus as operators on P we have

$$\lambda^* g(t) = \lambda^{-1} g(t) \lambda.$$

In view of the fact that $\lambda_f^* g(t) = g(\bar{f}(t))$, we get the following useful corollary.

Corollary 3.4.8   For any umbral operator λ_f we have

$$\lambda_f \circ g(\bar{f}(t)) = g(t) \circ \lambda_f \qquad (3.4.5)$$

for all $g(t)$ in F.

5. SHEFFER OPERATORS AND UMBRAL COMPOSITION

Let $s_n(x)$ be the Sheffer sequence for $(g(t) f(t))$. Then the linear operator λ on P defined by

$$\lambda x^n = s_n(x)$$

is called the *Sheffer operator* for $s_n(x)$ or for $(g(t), f(t))$. To indicate the dependence of λ on $g(t)$ and $f(t)$ we may write $\lambda_{g,f}$ or $\lambda_{g(t), f(t)}$. Clearly, Sheffer operators are bijective.

Since, if $s_n(x)$ is Sheffer for $(g(t), f(t))$, then $p_n(x) = g(t) s_n(x)$ is associated to $f(t)$, we see that

$$\lambda_{g,f} x^n = s_n(x)$$
$$= g(t)^{-1} p_n(x)$$
$$= g(t)^{-1} \lambda_f x^n,$$

and so

$$\lambda_{g,f} = g(t)^{-1} \lambda_f. \qquad (3.5.1)$$

This makes the theory of Sheffer operators a corollary of the theory of umbral operators. In particular, since $\lambda_{g,f}^* = \lambda_f^* (g(t)^{-1})^*$, we have

$$\lambda_{g,f}^* h(t) = g(\bar{f}(t))^{-1} h(\bar{f}(t)) \qquad (3.5.2)$$

for all $h(t)$ in $\mathscr{F}$.

Theorem 3.4.1 and Eq. (3.5.1) immediately give a characterization of Sheffer operators by their adjoints.

Theorem 3.5.1   A linear operator λ on P is a Sheffer operator if and only if its adjoint λ^* has the form of multiplication by an invertible series $g(t)^{-1}$ followed by an automorphism λ_f^* of $\mathscr{F}$. Moreover, if λ^* has this form, then $\lambda = \lambda_{g,f}$. In fact, any operator on $\mathscr{F}$ that is of this form is the adjoint of a Sheffer operator.

Let us consider the composition of Sheffer operators. If $\lambda_{g(t),\,f(t)}$ and $\lambda_{h(t),\,l(t)}$ are Sheffer operators, then by (3.5.2)

$$(\lambda_{g(t),\,f(t)} \circ \lambda_{h(t),\,l(t)})^* u(t) = \lambda_{h(t),\,l(t)}^* \circ \lambda_{g(t),\,f(t)}^* u(t)$$

$$= \lambda_{h(t),\,l(t)}^* g(\bar{f}(t))^{-1} u(\bar{f}(t))$$

$$= h(\bar{l}(t))^{-1} g(\bar{f}(\bar{l}(t)))^{-1} u(\bar{f}(\bar{l}(t)))$$

$$= (h \circ f)((l \circ f)(t))^{-1} g((l \circ f)(t))^{-1} u((l \circ f)(t))$$

$$= \lambda_{h(f(t))g(t),\,l(f(t))}^* u(t)$$

for all $u(t)$ in $\mathscr{F}$, and so

$$\lambda_{g(t),\,f(t)} \circ \lambda_{h(t),\,l(t)} = \lambda_{g(t)h(f(t)),\,l(f(t))}.$$

We also have by (3.5.1)

$$(\lambda_{g(t),\,f(t)}^{-1})^* u(t) = [\lambda_{f(t)}^{-1} \circ g(t)]^* u(t)$$

$$= g(t)^* (\lambda_{f(t)}^{-1})^* u(t)$$

$$= g(t) \lambda_{\bar{f}(t)}^* u(t)$$

$$= g(t) u(f(t))$$

$$= (g \circ \bar{f})(f(t)) u(f(t))$$

$$= \lambda_{g(\bar{f}(t))^{-1},\,\bar{f}(t)}^* u(t)$$

for all $u(t)$ in $\mathscr{F}$, and so

$$\lambda_{g(t),\,f(t)}^{-1} = \lambda_{g(\bar{f}(t))^{-1},\,\bar{f}(t)}.$$

We have proved the following theorem.

Theorem 3.5.2 The set of all Sheffer operators is a group under composition. In fact,

$$\lambda_{g(t),\,f(t)} \circ \lambda_{h(t),\,l(t)} = \lambda_{g(t)h(f(t)),\,l(f(t))},$$

$$\lambda_{g(t),\,f(t)}^{-1} = \lambda_{g(\bar{f}(t))^{-1},\,\bar{f}(t)}. \tag{3.5.3}$$

Corollary 3.5.3 A Sheffer operator maps Sheffer sequences to Sheffer sequences.

Proof Let $s_n(x)$ be a Sheffer sequence and let λ be a Sheffer operator. Then if $\mu: x^n \to s_n(x)$ is the Sheffer operator for $s_n(x)$, we have $\lambda s_n(x) = \lambda \circ \mu x^n$. But $\lambda \circ \mu$ is a Sheffer operator, and so $\lambda s_n(x)$ is a Sheffer sequence.

Corollary 3.5.4 If $s_n(x)$ and $r_n(x)$ are Sheffer sequences and $\alpha: s_n(x) \to r_n(x)$ is a linear operator, then α is a Sheffer operator.

Proof Let $\lambda: x^n \to s_n(x)$ and $\mu: x^n \to r_n(x)$ be Sheffer operators. Then $\mu \circ \lambda^{-1} s_n(x) = r_n(x) = \alpha s_n(x)$ and so $\alpha = \mu \circ \lambda^{-1}$ is a Sheffer operator.

Now we come to the behavior of Sheffer sequences under umbral composition.

Theorem 3.5.5 The set of Sheffer sequences is a group under umbral composition. In particular, if $s_n(x)$ is Sheffer for $(g(t), f(t))$ and $r_n(x)$ is Sheffer for $(h(t), l(t))$, then $r_n(s(x))$ is Sheffer for $(g(t)h(f(t)), l(f(t)))$. The identity under umbral composition is the sequence x^n and the inverse of the sequence $s_n(x)$ is the Sheffer sequence for $(g(\bar{f}(t))^{-1}, \bar{f}(t))$.

Proof From (3.5.3) we have

$$r_n(s(x)) = \lambda_{g(t), f(t)} r_n(x)$$

$$= \lambda_{g(t), f(t)} \circ \lambda_{h(t), l(t)} x^n$$

$$= \lambda_{g(t)h(f(t)), l(f(t))} x^n,$$

and this proves the second statement of the theorem. The third statement is clear. To prove the last statement we observe that $r_n(s(x)) = x^n$ if and only if $\lambda_{g(t)h(f(t)), l(f(t))}$ is the identity, which is true if and only if

$$g(t)h(f(t)) = 1,$$

$$l(f(t)) = t.$$

Solving these equations shows that $r_n(x)$ is Sheffer for

$$(h(t), l(t)) = (g(\bar{f}(t))^{-1}, \bar{f}(t)).$$

We summarize some of the previous results in a useful form in the next theorem.

Theorem 3.5.6 If $s_n(x)$ is Sheffer for $(g(t), f(t))$, then for all series $h(t)$ and polynomials $q(x)$

$$\langle h(t) | q(s(x)) \rangle = \langle g(\bar{f}(t))^{-1} h(\bar{f}(t)) | q(x) \rangle.$$

In particular,

$$\langle h(t) | s_n(x) \rangle = \langle g(\bar{f}(t))^{-1} h(\bar{f}(t)) | x^n \rangle.$$

If $p_n(x)$ is associated to $f(t)$, then

$$\langle h(t) | q(\mathbf{p}(x)) \rangle = \langle h(\bar{f}(t)) | q(x) \rangle$$

and

$$\langle h(t) | p_n(x) \rangle = \langle h(\bar{f}(t)) | x^n \rangle.$$

6. UMBRAL SHIFTS AND THE RECURRENCE FORMULA FOR ASSOCIATED SEQUENCES

Let $p_n(x)$ be the associated sequence for $f(t)$. Then the linear operator θ on P defined by

$$\theta p_n(x) = p_{n+1}(x)$$

is called the *umbral shift* for $p_n(x)$ or for $f(t)$. We use the notation θ_f to indicate the dependence on $f(t)$.

In the next section we shall study the more general *Sheffer shift*

$$\theta: s_n(x) \to s_{n+1}(x)$$

for $s_n(x)$ an arbitrary Sheffer sequence.

Since $\deg p_n(x) = n$, we see that an umbral shift is injective, but it is not surjective.

We proceed to characterize umbral shifts by means of their adjoints. If $\theta_f: p_n(x) \to p_{n+1}(x)$ is an umbral shift, we have

$$
\begin{aligned}
\langle \theta_f^* f(t)^k \mid p_n(x) \rangle &= \langle f(t)^k \mid \theta_f p_n(x) \rangle \\
&= \langle f(t)^k \mid p_{n+1}(x) \rangle \\
&= (n+1)! \, \delta_{n+1,k} \\
&= kn! \, \delta_{n,k+1} \\
&= \langle kf(t)^{k-1} \mid p_n(x) \rangle
\end{aligned}
$$

for all $k \geq 0$, and so

$$\theta_f^* f(t)^k = kf(t)^{k-1}. \tag{3.6.1}$$

By the expansion theorem any $g(t)$ in $\mathscr{F}$ can be written in the form

$$g(t) = \sum_{k=0}^{\infty} a_k f(t)^k,$$

and the continuity of θ_f^* implies

$$\theta_f^* g(t) = \sum_{k=0}^{\infty} ka_k f(t)^{k-1}.$$

Also, we have

$$
\begin{aligned}
\theta_f^* f(t)^k f(t)^j &= \theta_f^* f(t)^{k+j} \\
&= (k+j) f(t)^{k+j-1} \\
&= jf(t)^k f(t)^{j-1} + kf(t)^{k-1} f(t)^j \\
&= f(t)^k \theta_f^* f(t)^j + f(t)^j \theta_f^* f(t)^k.
\end{aligned}
$$

From this and the continuity of θ_f^* we get

$$\theta_f^* g(t)h(t) = g(t)\theta_f^* h(t) + h(t)\theta_f^* g(t)$$

for all $g(t)$ and $h(t)$ in $\mathscr{F}$.

Thus θ_f^* is a surjective derivation on $\mathscr{F}$. In view of (3.6.1), the map θ_f^* is actually the derivative with respect to $f(t)$, which we denote by ∂_f. That is, $\theta_f^* = \partial_f$. Once again we have a characterization.

Theorem 3.6.1 A linear operator θ on P is an umbral shift if and only if its adjoint θ^* is a surjective derivation on $\mathscr{F}$. Moreover, if θ_f is an umbral shift, then $\theta_f^* = \partial_f$ is the derivative with respect to $f(t)$,

$$\theta_f^* f(t)^k = kf(t)^{k-1}$$

for all $k \geq 0$. In particular, $\theta_f^* f(t) = 1$.

Proof We have already established one half of this theorem. Suppose now that θ^* is a surjective derivation on $\mathscr{F}$. It follows from Theorem 3.2.3 that there exists a delta functional $f(t)$ for which $\theta^* f(t) = 1$. If $p_n(x)$ is associated to $f(t)$, then

$$
\begin{aligned}
\langle f(t)^k \,|\, \theta p_n(x) \rangle &= \langle \theta^* f(t)^k \,|\, p_n(x) \rangle \\
&= \langle kf(t)^{k-1}\theta^* f(t) \,|\, p_n(x) \rangle \\
&= \langle kf(t)^{k-1} \,|\, p_n(x) \rangle \\
&= (n+1)!\,\delta_{n+1,k} \\
&= \langle f(t)^k \,|\, p_{n+1}(x) \rangle,
\end{aligned}
$$

and so $\theta p_n(x) = p_{n+1}(x)$ and θ is the umbral shift for $f(t)$.

Corollary 3.2.4, along with Theorems 3.3.1 and 3.6.1, implies that any surjective derivation on $\mathscr{F}$ is the adjoint of an umbral shift.

Theorem 3.6.2 The adjoint map, which sends a linear operator θ on P to its adjoint operator θ^* on $\mathscr{F}$, is a bijection between the set of all umbral shifts on P and the set of all surjective derivations on $\mathscr{F}$.

We can now fill in the remaining box of our original diagram (see Fig. 3.4).

Unlike the case for automorphisms, the set of surjective derivations on $\mathscr{F}$ does not form a group under composition. In casting about for some structure, we recall a well-known fact that almost every student of elementary calculus fails to learn, namely, any two derivations are related by a formula known as the chain rule!

Theorem 3.6.3 (The Chain Rule) Let ∂_f and ∂_g be surjective derivations on $\mathscr{F}$. Then

$$\partial_g = \left(\partial_g f(t)\right)\partial_f.$$

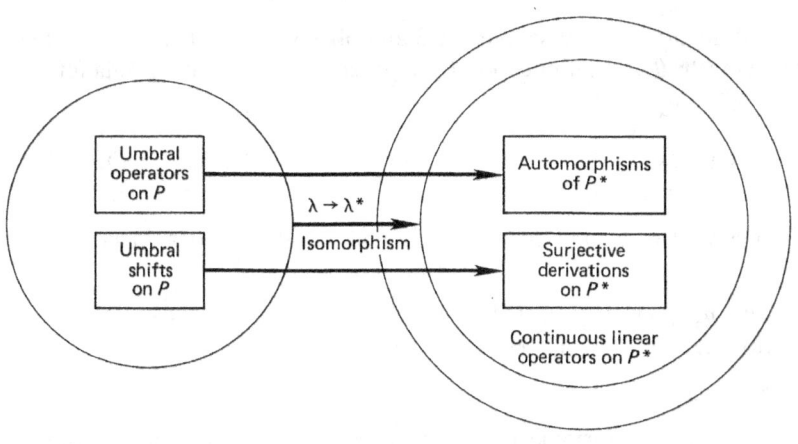

Linear operators on P Linear operators on P^*

FIGURE 3.4

Proof We have

$$\partial_g f(t)^k = kf(t)^{k-1}\partial_g f(t) = \left(\partial_g f(t)\right)\partial_f f(t)^k$$

for all $k \geq 0$ and the result follows by continuity.

Corollary 3.6.4 If ∂_f and ∂_g are surjective derivations on $\mathscr{F}$, then

$$\partial_g f(t) = \frac{1}{\partial_f g(t)}.$$

The correspondence of Theorem 3.6.2 can be used to translate the chain rule into a relationship between umbral shifts.

Theorem 3.6.5 Let θ_f and θ_g be umbral shifts. Then

$$\theta_f = \theta_g \circ \left(\partial_f g(t)\right).$$

Proof According to the chain rule, we have

$$\theta_f^* = \left(\partial_f g(t)\right)\theta_g^*,$$

and so

$$\begin{aligned}
\langle h(t) \,|\, \theta_f p(x)\rangle &= \langle \theta_f^* h(t) \,|\, p(x)\rangle \\
&= \langle (\partial_f g(t))\theta_g^* h(t) \,|\, p(x)\rangle \\
&= \langle \theta_g^* h(t) \,|\, (\partial_f g(t))p(x)\rangle \\
&= \langle h(t) \,|\, \theta_g \circ (\partial_f g(t))p(x)\rangle
\end{aligned}$$

for all $h(t)$ in $\mathscr{F}$ and $p(x)$ in P. Thus $\theta_f = \theta_g \circ \left(\partial_f g(t)\right)$.

Taking $g(t) = t$ in Theorem 3.6.5 and observing that $\theta_g x^n = x^{n+1}$, that is, the operator θ_g is multiplication by x, we get a very useful formula for θ_f,

$$\theta_f = x \, \partial_f t = x(\partial_t f(t))^{-1} = x(f'(t))^{-1}.$$

Applying this to the associated sequence for $f(t)$ leads to the following recurrence formulas.

Corollary 3.6.6 (The Recurrence Formulas) If $p_n(x)$ is associated to $f(t)$, then

(i) $p_{n+1}(x) = x[f'(t)]^{-1} p_n(x)$ and
(ii) $p_{n+1}(x) = x\lambda_f [\bar{f}(t)]' x^n$

for all $n \geq 0$.

Proof The first formula has been proved. For the second we use Corollary 3.4.8,

$$
\begin{aligned}
p_{n+1}(x) &= x[f'(t)]^{-1} \lambda_f x^n \\
&= x\lambda_f [f'(\bar{f}(t))]^{-1} x^n \\
&= x\lambda_f [\bar{f}(t)]' x^n.
\end{aligned}
$$

Incidentally, the second form of the recurrence formula can be written as

$$\lambda_f x^{n+1} = x\lambda_f [\bar{f}(t)]' x^n,$$

which is equivalent to the commutation rule

$$\lambda_f x = x\lambda_f [\bar{f}(t)]'.$$

We can now derive a formula for surjective derivations on $\mathscr{F}$.

Theorem 3.6.7 Let θ be an umbral shift. Then as operators on p we have

$$\theta^* g(t) = g(t)\theta - \theta g(t) \tag{3.6.2}$$

for all operators $g(t)$ in $\mathscr{F}$.

Proof For any $f(t)$ in $\mathscr{F}$ and $p(x)$ in P we have

$$
\begin{aligned}
\langle f(t) \,|\, (\theta^* g(t)) p(x) \rangle &= \langle f(t)(\theta^* g(t)) \,|\, p(x) \rangle \\
&= \langle \theta^* f(t) g(t) - g(t)\theta^* f(t) \,|\, p(x) \rangle \\
&= \langle \theta^* f(t) g(t) \,|\, p(x) \rangle - \langle g(t)\theta^* f(t) \,|\, p(x) \rangle \\
&= \langle f(t) \,|\, g(t)\theta p(x) \rangle - \langle f(t) \,|\, \theta g(t) p(x) \rangle \\
&= \langle f(t) \,|\, [g(t)\theta - \theta g(t)] p(x) \rangle,
\end{aligned}
$$

from which the desired conclusion follows.

When $\theta: x^n \to x^{n+1}$ is the umbral shift for $p_n(x) = x^n$, the map θ^* is ∂_t and (3.6.2) becomes

$$g'(t) = g(t)x - xg(t), \tag{3.6.3}$$

where both sides are considered as operators on P. The right side of this equation is known as the Pincherle derivative of the operator $g(t)$.

7. SHEFFER SHIFTS AND THE RECURRENCE FORMULA FOR SHEFFER SEQUENCES

Let $s_n(x)$ be the Sheffer sequence for $(g(t), f(t))$. The linear operator θ on P defined by

$$\theta s_n(x) = s_{n+1}(x) \tag{3.7.1}$$

is called the *Sheffer shift* for $s_n(x)$ or for $(g(t), f(t))$. We also use the notation $\theta_{f,g}$. Since $\deg s_n(x) = n$, a Sheffer shift is injective, but it is not surjective.

If $s_n(x)$ is Sheffer for $(g(t), f(t))$, then $p_n(x) = g(t)s_n(x)$ is associated to $f(t)$. and so (3.7.1) becomes

$$\theta_{g,f} g(t)^{-1} p_n(x) = g(t)^{-1} p_{n+1}(x).$$

Thus

$$\theta_{g,f} = g(t)^{-1} \theta_f g(t),$$

where θ_f is the umbral shift for $f(t)$. Using Theorems 3.6.6 and 3.6.7 and the chain rule, we get

$$\begin{aligned}
\theta_{g,f} &= g(t)^{-1} \theta_f g(t) \\
&= g(t)^{-1} [g(t)\theta_f - \theta_f^* g(t)] \\
&= \theta_f - g(t)^{-1} \partial_{f(t)} g(t) \\
&= \theta_f - g(t)^{-1} \partial_{f(t)} t \, \partial_t g(t) \\
&= x[f'(t)]^{-1} - g(t)^{-1}[f'(t)]^{-1} g'(t) \\
&= \left[x - \frac{g'(t)}{g(t)} \right] \frac{1}{f'(t)}.
\end{aligned}$$

We have proved the following result.

Theorem 3.7.1 If $\theta_{g,f}$ is a Sheffer shift, then

$$\theta_{g,f} = \left[x - \frac{g'(t)}{g(t)} \right] \frac{1}{f'(t)}.$$

This gives a recurrence formula for Sheffer sequences.

Corollary 3.7.2 (The Recurrence Formula for Sheffer Sequences) If $s_n(x)$ is Sheffer for $(g(t), f(t))$, then

$$s_{n+1}(x) = \left[x - \frac{g'(t)}{g(t)} \right] \frac{1}{f'(t)} s_n(x).$$

8. THE TRANSFER FORMULAS

We next derive two powerful formulas for the computation of an associated sequence from its delta functional. These formulas, together with Theorem 2.3.6, can be used to compute Sheffer sequences.

Theorem 3.8.1 (The Transfer Formulas) If $p_n(x)$ is the associated sequence for $f(t)$, then

(i) $p_n(x) = f'(t)\left(\dfrac{f(t)}{t}\right)^{-n-1} x^n$ and

(ii) $p_n(x) = x\left(\dfrac{f(t)}{t}\right)^{-n} x^{n-1}$

for all $n \geq 0$.

Proof For convenience we set $g(t) = f(t)/t$. Then $g(t)$ is invertible. Actually, the right sides of (i) and (ii) can be shown to be equal, using (3.6.3),

$$\begin{aligned}
f'(t)g(t)^{-n-1}x^n &= [tg(t)]'g(t)^{-n-1}x^n \\
&= g(t)^{-n}x^n + tg'(t)g(t)^{-n-1}x^n \\
&= g(t)^{-n}x^n + ng'(t)g(t)^{-n-1}x^{n-1} \\
&= g(t)^{-n}x^n - [g(t)^{-n}]'x^{n-1} \\
&= g(t)^{-n}x^n - [g(t)^{-n}x - xg(t)^{-n}]x^{n-1} \\
&= xg(t)^{-n}x^{n-1}.
\end{aligned}$$

To prove (i) we verify the two conditions of Theorem 2.4.5 for the sequence $q_n(x) = f'(t)g(t)^{-n-1}x^n$. For the first condition, when $n \geq 1$, the fourth equality above and Theorem 2.1.10 give

$$\begin{aligned}
\langle t^0 \,|\, q_n(x) \rangle &= \langle t^0 \,|\, f'(t)g(t)^{-n-1}x^n \rangle \\
&= \langle t^0 \,|\, g(t)^{-n}x^n - [g(t)^{-n}]'x^{n-1} \rangle \\
&= \langle g(t)^{-n} \,|\, x^n \rangle - \langle [g(t)^{-n}]' \,|\, x^{n-1} \rangle \\
&= \langle g(t)^{-n}x^n \rangle - \langle g(t)^{-n} \,|\, x^n \rangle \\
&= 0.
\end{aligned}$$

If $n = 0$, then $\langle t^0 \mid q_n(x) \rangle = 1$, and so $\langle t^0 \mid q_n(x) \rangle = \delta_{n,0}$. For the second condition

$$
\begin{aligned}
f(t)q_n(x) &= f(t)f'(t)g(t)^{-n-1}x^n \\
&= tg(t)f'(t)g(t)^{-n-1}x^n \\
&= nf'(t)g(t)^{-n}x^{n-1} \\
&= nq_{n-1}(x).
\end{aligned}
$$

Thus $q_n(x)$ is associated to $f(t)$ and the proof is complete.

The next corollary gives a formula that directly relates any two associated sequences.

Corollary 3.8.2 Let $p_n(x)$ be associated to $f(t)$ and let $q_n(x)$ be associated to $g(t)$. Then for $n \geq 1$,

$$
q_n(x) = x \left(\frac{f(t)}{g(t)} \right)^n x^{-1} p_n(x).
$$

[We remark that $f(t)/g(t)$ is the invertible element of $\mathscr{F}$ defined by $[f(t)/t][g(t)/t]^{-1}$ and that since $p_n(0) = 0$ for $n \geq 1$, the expression $x^{-1}p_n(x)$ is indeed a polynomial.]

Proof By the first transfer formula we have

$$
(f(t)/t)^n x^{-1} p_n(x) = x^{n-1} = (g(t)/t)^n x^{-1} q_n(x),
$$

and the result follows by solving for $q_n(x)$.

We conclude this section with a rather useful result.

Theorem 3.8.3 If $s_n(x)$ is Sheffer for $(g(t), f(t))$, then

$$
r_n(x) = s_n(x + \alpha + \beta n)
$$

is Sheffer for

$$
\left(g(t)e^{-\alpha t}\left(1 - \frac{\beta f(t)}{f'(t)} \right), e^{-\beta t}f(t) \right).
$$

Proof The sequence $r_n(x)$ is Sheffer for the delta functional $e^{-\beta t}f(t)$ since

$$
\begin{aligned}
e^{-\beta t}f(t)r_n(x) &= e^{-\beta t}f(t)e^{(\alpha + \beta n)t}s_n(x) \\
&= ne^{(\alpha + \beta n - \beta)t}s_{n-1}(x) \\
&= nr_{n-1}(x).
\end{aligned}
$$

Now suppose that $r_n(x)$ is Sheffer for $(h(t), e^{-\beta t}f(t))$. Corollary 3.8.2 implies that the associated sequence for $e^{-\beta t}f(t)$ is

$$q_n(x) = xe^{\beta nt}x^{-1}p_n(x)$$

$$= \frac{x}{x + \beta n}p_n(x + \beta n)$$

$$= e^{\beta nt}\left(\frac{x - \beta n}{x}\right)p_n(x),$$

and so

$$h(t)s_n(x + \alpha + \beta n) = e^{\beta nt}\left(\frac{x - \beta n}{x}\right)p_n(x)$$

or

$$h(t)s_n(x) = e^{-\alpha t}\left(\frac{x - \beta n}{x}\right)p_n(x)$$

$$= e^{-\alpha t}(1 - \beta nx^{-1})p_n(x).$$

By the recurrence formula, $p_n(x) = x[f'(t)]^{-1}p_{n-1}(x)$, and so

$$h(t)s_n(x) = e^{-\alpha t}[p_n(x) - \beta n(f'(t))^{-1}p_{n-1}(x)]$$

$$= e^{-\alpha t}[p_n(x) - \beta f(t)(f'(t))^{-1}p_n(x)]$$

$$= e^{-\alpha t}\left(1 - \frac{\beta f(t)}{f'(t)}\right)p_n(x)$$

$$= e^{-\alpha t}\left(1 - \frac{\beta f(t)}{f'(t)}\right)g(t)s_n(x).$$

This gives the desired result.

CHAPTER 4

EXAMPLES

In this chapter we discuss some important examples of Sheffer sequences. It is not our intention here to give a treatise on any polynomial sequence, but rather to give enough examples of umbral techniques to allow the interested reader to continue on his own. In the next chapter we shall develop additional umbral techniques and apply them to the examples of this chapter.

Our primary source for the definitions of special polynomial sequences is the Bateman Manuscript Project (Erdelyi [1]), and we remind the reader of the remarks made in this regard following Theorem 2.3.4. We shall keep the reader informed of any discrepancies, albeit they are only minor, between our definitions and those of our sources.

In the hope of achieving greater clarity each example is divided into five parts, whose defining lines are by no means precise:

definitions,
the generating function and the Sheffer identity,
recurrence relations and operational formulas,
expansions, and
miscellaneous results.

1. ASSOCIATED SEQUENCES

For ease of reference we list the key results. Let the sequence $p_n(x)$ be associated to $f(t)$.

Generating function

$$\sum_{k=0}^{\infty} \frac{p_k(x)}{k!} t^k = e^{x \bar{f}(t)}.$$

Binomial identity

$$p_n(x + y) = \sum_{k=0}^{n} \binom{n}{k} p_k(x) p_{n-k}(y).$$

Theorem 2.4.5

$$\langle t^0 \mid p_n(x) \rangle = \delta_{n, 0},$$

$$f(t) p_n(x) = n p_{n-1}(x).$$

Recurrence formulas

$$p_{n+1}(x) = x[f'(t)]^{-1} p_n(x),$$

$$p_{n+1}(x) = x \lambda_f [\bar{f}(t)]' x^n.$$

Transfer formulas

$$p_n(x) = x \left(\frac{f(t)}{t} \right)^{-n} x^{n-1},$$

$$p_n(x) = f'(t) \left(\frac{f(t)}{t} \right)^{-n-1} x^n.$$

Expansion theorem

$$g(t) = \sum_{k=0}^{\infty} \frac{\langle g(t) \mid p_k(x) \rangle}{k!} f(t)^k.$$

Polynomial expansion theorem

$$p(x) = \sum_{k \geq 0} \frac{\langle f(t)^k \mid p(x) \rangle}{k!} p_k(x).$$

Theorem 3.5.6

$$\langle g(t) \mid q(\mathbf{p}(x)) \rangle = \langle g(\bar{f}(t)) \mid q(x) \rangle,$$

$$\langle g(t) \mid p_n(x) \rangle = \langle g(\bar{f}(t)) \mid x^n \rangle.$$

Theorem 2.4.8

$$x p_n(x) = \sum_{k=1}^{n+1} \binom{n}{k-1} \langle f'(t) \mid p_{n-k+1}(x) \rangle p_k(x)$$

or

$$x p_n(x) = \sum_{k=1}^{n+1} \binom{n}{k-1} \langle f'(\bar{f}(t)) \mid x^{n-k+1} \rangle p_k(x).$$

Theorem 2.4.9

$$p_n'(x) = \sum_{k=0}^{n-1} \binom{n}{k} \langle t \mid p_{n-k}(x) \rangle p_k(x)$$

or

$$p_n'(x) = \sum_{k=0}^{n-1} \binom{n}{k} \langle \bar{f}(t) \mid x^{n-k} \rangle p_k(x).$$

Conjugate representation

$$p_n(x) = \sum_{k=1}^{n} \frac{\langle \bar{f}(t)^k \mid x^n \rangle}{k!} x^k.$$

Proposition 2.1.11

$$\langle g(t) \mid p(ax) \rangle = \langle g(at) \mid p(x) \rangle.$$

Theorem 2.1.10

$$\langle g(t) \mid xp(x) \rangle = \langle \partial_t g(t) \mid p(x) \rangle.$$

1.1. The Sequence x^n

The sequence $p_n(x) = x^n$ is, of course, associated to $f(t) = t$. The results in this case present no surprise and we list them quickly. The generating function is

$$\sum_{k=0}^{\infty} \frac{x^k}{k!} t^k = e^{xt},$$

and the binomial identity is

$$(x + y)^n = \sum_{k=0}^{n} \binom{n}{k} x^k y^{n-k}.$$

The expansion theorem is

$$g(t) = \sum_{k=0}^{\infty} \frac{\langle g(t) \mid x^k \rangle}{k!} t^k,$$

which for $g(t) = e^{yt}$ is Taylor's expansion

$$e^{yt} = \sum_{k=0}^{\infty} \binom{y}{k} t^k.$$

Applying this to $p(x)$ gives

$$p(x + y) = \sum_{k \geq 0} \binom{y}{k} p^{(k)}(x).$$

1.2. The Lower Factorial Polynomials

DEFINITIONS

Let us consider the forward difference functional

$$f(t) = e^{at} - 1$$

for $a \neq 0$. We note that

$$f'(t) = ae^{at},$$
$$\bar{f}(t) = a^{-1}\log(1 + t).$$

The associated sequence for $f(t)$ is easily determined by using Theorem 3.5.6,

$$
\begin{aligned}
p_n(y) &= \langle e^{yt} \mid p_n(x) \rangle \\
&= \langle e^{y\bar{f}(t)} \mid x^n \rangle \\
&= \langle e^{ya^{-1}\log(1 + t)} \mid x^n \rangle \\
&= \langle (1 + t)^{y/a} \mid x^n \rangle \\
&= \left\langle \sum_{k=0}^{\infty} \binom{y/a}{k} t^k \,\middle|\, x^n \right\rangle \\
&= \left(\frac{y}{a} \right)_n.
\end{aligned}
$$

Thus

$$p_n(x) = \left(\frac{x}{a}\right)_n = \frac{x}{a}\left(\frac{x}{a} - 1\right)\cdots\left(\frac{x}{a} - n + 1\right).$$

These are the lower factorial polynomials, also known as the falling factorial or binomial polynomials.

The first form of the recurrence formula would have worked as well to determine the sequence $p_n(x)$,

$$
\begin{aligned}
p_{n+1}(x) &= x[f'(t)]^{-1}p_n(x) \\
&= a^{-1}xe^{-at}p_n(x) \\
&= \frac{x}{a}p_n(x - a),
\end{aligned}
$$

and since $p_0(x) = 1$, we are lead to $p_n(x) = (x/a)_n$.

For ease of expression we couch some of the following results for $a = 1$, although it is really no more difficult to obtain them for arbitrary values of $a \neq 0$.

We express $(x)_n$ in powers of x by

$$(x)_n = \sum_{k=0}^{n} \frac{\langle t^k \mid (x)_n \rangle}{k!} x^k.$$

The coefficients in this expression are known as the Stirling numbers of the first kind,

$$s(n, k) = \frac{1}{k!} \langle t^k \mid (x)_n \rangle.$$

Thus

$$(x)_n = \sum_{k=0}^{n} s(n, k) x^k.$$

We shall discuss these important numbers later in this example and in the next one.

GENERATING FUNCTION AND SHEFFER IDENTITY

The generating function for $(x)_n$ is

$$\sum_{k=0}^{\infty} \frac{(x)_k}{k!} t^k = e^{x \log(1 + t)},$$

which can be written in the familiar form

$$\sum_{k=0}^{\infty} \binom{x}{k} t^k = (1 + t)^x.$$

The binomial identity is

$$(x + y)_n = \sum_{k=0}^{n} \binom{n}{k} x_k y_{n-k},$$

which can also be written as

$$\binom{x + y}{n} = \sum_{k=0}^{n} \binom{x}{k} \binom{y}{n - k}.$$

This is known as the Vandermonde convolution formula.

RECURRENCE RELATIONS AND OPERATIONAL FORMULAS

Theorem 2.4.5 gives the formula

$$(e^t - 1)(x)_n = n(x)_{n-1},$$

which can be written as

$$(x+1)_n - (x)_n = n(x)_{n-1}.$$

The second form of the recurrence formula is

$$\begin{aligned}
(x)_{n+1} &= x\lambda_f[\bar{f}(t)]'x^n \\
&= x\lambda_f(1+t)^{-1}x^n \\
&= x\lambda_f \sum_{k=0}^{n} (-1)^k t^k x^n \\
&= x \sum_{k=0}^{n} (-1)^k (n)_k (x)_{n-k} \\
&= x \sum_{k=0}^{n} (-1)^{n-k} (n)_{n-k} (x)_k.
\end{aligned}$$

EXPANSIONS

The expansion theorem is

$$g(t) = \sum_{k=0}^{\infty} \frac{\langle g(t) | (x/a)_k \rangle}{k!} (e^{at} - 1)^k.$$

Taking $g(t) = e^{yt}$ we get the beautiful formula

$$e^{yt} = \sum_{k=0}^{\infty} \binom{y/a}{k} (e^{at} - 1)^k.$$

This is actually Newton's forward difference interpolation formula. If we use the notation

$$E^y : p(x) \to p(x+y),$$
$$\Delta_a : p(x) \to p(x+a) - p(x)$$

for the translation and forward difference operators, it takes the more familiar form

$$E^y = \sum_{k=0}^{\infty} \binom{y/a}{k} \Delta_a^k.$$

Setting $f(t) = t^m$ in the expansion theorem gives

$$\begin{aligned}
t^m &= \sum_{k=0}^{\infty} \frac{\langle t^m | (x/a)_k \rangle}{k!} (e^{at} - 1)^k \\
&= \sum_{k=0}^{\infty} \frac{m!}{k! a^m} s(k, m)(e^{at} - 1)^k.
\end{aligned}$$

By applying this to a polynomial $p(x)$ and observing that

$$\langle(e^{at} - 1)^k \,|\, p(x)\rangle = \sum_{j=0}^{k}\binom{k}{j}(-1)^{k-j}\langle e^{jat} \,|\, p(x)\rangle$$

$$= \sum_{j=0}^{k}\binom{k}{j}(-1)^{k-j}p(ja), \qquad (4.1.1)$$

we obtain a classical formula for numerical differentiation,

$$p^{(m)}(0) = \sum_{k\geq 0}\frac{m!}{k!a^m}s(k,m)\sum_{j=0}^{k}\binom{k}{j}(-1)^{k-j}p(ja).$$

Expanding the functional $(e^{yt} - 1)/t$ in terms of the forward difference gives

$$\frac{e^{yt} - 1}{t} = \sum_{k=0}^{\infty}\frac{\langle(e^{yt} - 1)/t \,|\, (x)_k\rangle}{k!}(e^t - 1)^k.$$

If we apply this to a polynomial $p(x)$, we get

$$\int_{0}^{y}p(u)\,du = \sum_{k\geq 0}\frac{\langle(e^{yt} - 1)/t \,|\, (x)_k\rangle}{k!}(e^t - 1)^k p(x).$$

This is known as Gregory's formula (Jordan [1, p. 284]). It was discovered by Gregory in 1670 and is reported to be the earliest formula of numerical integration. We shall discuss the coefficients of this expansion when we come to the Bernoulli polynomials of the second kind in Section 3.

The polynomial expansion theorem is

$$p(x) = \sum_{k\geq 0}\frac{\langle(e^{at} - 1)^k \,|\, p(x)\rangle}{k!}\left(\frac{x}{a}\right)_k, \qquad (4.1.2)$$

and, using (4.1.1), this becomes

$$p(x) = \sum_{k\geq 0}\binom{x/a}{k}\sum_{j=0}^{k}\binom{k}{j}(-1)^{k-j}p(ja).$$

If we take $a = 1$ and $p(x) = x^n$ in (4.1.2), we obtain

$$x^n = \sum_{k=0}^{n}\frac{\langle(e^t - 1)^k \,|\, x^n\rangle}{k!}x^k.$$

These coefficients are known as the Stirling numbers of the second kind,

$$S(n, k) = \frac{1}{k!}\langle(e^t - 1)^k \,|\, x^n\rangle.$$

Thus

$$x^n = \sum_{k=0}^{n} S(n,k)(x)_k. \tag{4.1.3}$$

We shall say more about these numbers later, but for now we notice that (4.1.1) gives

$$S(n,k) = \frac{1}{k!} \sum_{j=0}^{k} \binom{k}{j}(-1)^{k-j} j^n.$$

Again, by the polynomial expansion theorem

$$\left(\frac{x}{b}\right)_n = \sum_{k=0}^{n} \frac{1}{k!} \langle (e^{at} - 1)^k \,|\, (x/b)_n \rangle \left(\frac{x}{a}\right)_k.$$

From Theorem 3.5.6 we get

$$\langle (e^{at} - 1)^k \,|\, (x/b)_n \rangle = \langle [e^{(a/b)\log(1+t)} - 1]^k \,|\, x^n \rangle$$
$$= \langle (1+t)^{a/b} - 1]^k \,|\, x^n \rangle,$$

and, using Proposition 2.1.3, we have

$$\langle (e^{at} - 1)^k \,|\, (x/b)_n \rangle$$

$$= \sum_{i_1 + \cdots + i_k = n} \binom{n}{i_1, \ldots, i_k} \langle (1+t)^{a/b} - 1 \,|\, x^{i_1} \rangle \cdots \langle (1+t)^{a/b} - 1 \,|\, x^{i_k} \rangle$$

$$= \sum_{\substack{i_1 + \cdots + i_k = n \\ i_j > 0}} \binom{n}{i_1, \ldots, i_k} \left(\frac{a}{b}\right)_{i_1} \cdots \left(\frac{a}{b}\right)_{i_k}.$$

Either of these forms can be substituted into the above expansion.
    According to Theorem 2.4.8,

$$x\left(\frac{x}{a}\right)_n = \sum_{k=1}^{n+1} \binom{n}{k-1} \langle f'(t) \,|\, (x/a)_{n-k+1} \rangle \left(\frac{x}{a}\right)_k$$

$$= \sum_{k=1}^{n+1} \binom{n}{k-1} \langle ae^{at} \,|\, (x/a)_{n-k+1} \rangle \left(\frac{x}{a}\right)_k$$

$$= a \sum_{k=1}^{n+1} \binom{n}{k-1} (1)_{n-k+1} \left(\frac{x}{a}\right)_k,$$

and since $(1)_m = \delta_{m,0} + \delta_{m,1}$, we get

$$x\left(\frac{x}{a}\right)_n = a\left(\frac{x}{a}\right)_{n+1} + an\left(\frac{x}{a}\right)_n.$$

Theorem 2.4.9 provides a formula for the derivative of the lower factorial polynomials,

$$t(x)_n = \sum_{k=0}^{n-1}\binom{n}{k}(-1)^{n-k}(n-k-1)!\,(x)_k.$$

Miscellaneous Results

Let us explore some of the properties of the Stirling numbers, using umbral methods. A great deal has been written about these numbers, and they have important combinatorial significance which we cannot discuss here. The interested reader might wish to consult Comtet [1] or Riordan [2, 3].

As to the Stirling numbers of the first kind

$$s(n,k) = \frac{1}{k!}\langle t^k \mid (x)_n \rangle,$$

let us apply $t^k / k!$ to both sides of the recurrence

$$x(x)_n = (x)_{n+1} + n(x)_n.$$

Using Theorem 2.1.10, we get

$$\frac{1}{k!}\langle t^k \mid x(x)_n \rangle = \frac{1}{(k-1)!}\langle t^{k-1} \mid (x)_n \rangle = s(n,k-1)$$

and

$$\frac{1}{k!}\langle t^k \mid (x)_{n+1} + n(x)_n \rangle = s(n+1,k) + ns(n,k),$$

and so

$$s(n+1,k) = s(n,k-1) - ns(n,k).$$

This recurrence, together with the initial conditions

$$s(n,0) = 0, \qquad n > 0,$$
$$s(0,k) = 0, \qquad k > 0,$$
$$s(0,0) = 0,$$

enables one to compute the numbers $s(n,k)$.

Applying $t^k / k!$ to the formula obtained from the second form of the recurrence formula gives

$$s(n+1,k) = \sum_{j=0}^{n}(-1)^{n-j}(n)_{n-j}s(j,k-1).$$

The conjugate representation for $(x)_n$ is

$$(x)_n = \sum_{k=0}^{n}\frac{1}{k!}\langle [\log(1+t)]^k \mid x^n \rangle x^k,$$

and so

$$s(n, k) = \frac{1}{k!} \langle [\log(1 + t)]^k \,|\, x^n \rangle.$$

From Proposition 2.1.3 we then get

$$
\begin{aligned}
s(n, k) &= \frac{1}{k!} \sum_{i_1 + \cdots + i_k = n} \binom{n}{i_1, \ldots, i_k} \langle \log(1 + t) \,|\, x^{i_1} \rangle \cdots \langle \log(1 + t) \,|\, x^{i_k} \rangle \\
&= \frac{1}{k!} \sum_{\substack{i_1 + \cdots + i_k = n \\ i_j > 0}} \binom{n}{i_1, \ldots, i_k} (-1)^{n-k} \left(\frac{i_1!}{i_1} \right) \cdots \left(\frac{i_k!}{i_k} \right) \\
&= (-1)^{n-k} \frac{n!}{k!} \sum_{\substack{i_1 + \cdots + i_k = n \\ i_j > 0}} \frac{1}{i_1 \cdots i_k}.
\end{aligned}
$$

When the binomial identity for $(x)_n$ is written out and the coefficients of like powers of x and y are compared, one obtains

$$\binom{i + j}{i} s(n, i + j) = \sum_{k=0}^{n} \binom{n}{k} s(k, i) s(n - k, j).$$

Turning to the Stirling numbers of the second kind, we have already remarked that

$$S(n, k) = \frac{1}{k!} \sum_{j=0}^{k} \binom{k}{j} (-1)^{k-j} j^n.$$

Proposition 2.1.3 gives, as we have seen,

$$
\begin{aligned}
S(n, k) &= \frac{1}{k!} \langle (e^t - 1)^k \,|\, x^n \rangle \\
&= \frac{1}{k!} \sum_{\substack{i_1 + \cdots + i_k = n \\ i_j > 0}} \binom{n}{i_1, \ldots, i_k}.
\end{aligned}
$$

Finally, for $k \geq 1$ and $n \geq 1$, we have from Theorem 2.1.10

$$
\begin{aligned}
S(n, k) &= \frac{1}{k!} \langle (e^t - 1)^k \,|\, x^n \rangle \\
&= \frac{1}{(k-1)!} \langle e^t (e^t - 1)^{k-1} \,|\, x^{n-1} \rangle \\
&= \frac{1}{(k-1)!} \langle (e^t - 1)^k \,|\, x^{n-1} \rangle + \frac{1}{(k-1)!} \langle (e^t - 1)^{k-1} \,|\, x^{n-1} \rangle \\
&= k S(n - 1, k) + S(n - 1, k - 1).
\end{aligned}
$$

From the initial conditions

$$S(n,0) = 0, \qquad n > 0,$$
$$S(0,k) = 0, \qquad k > 0,$$
$$S(0,0) = 1,$$

one can compute the numbers $S(n, k)$.

We shall discuss the connection between the Stirling numbers of the first and second kind in the next example.

The multinomial version of Theorem 2.3.10 and Eq. (4.1.1) lead to

$$\sum_{j=0}^{k} \binom{k}{j} (-1)^{k-j} p_n(ja) = \sum_{\substack{i_1 + \cdots + i_k = n \\ i_j > 0}} \binom{n}{i_1, \ldots, i_k} p_{i_1}(a) \cdots p_{i_k}(a) \quad (4.1.4)$$

for any associated sequence. This specializes to give some interesting combinatorial identities. For example, if we take $p_n(x) = (bx/a)_n$, this becomes

$$\sum_{j=0}^{k} \binom{k}{j} (-1)^{k-j} \binom{bj}{n} = \sum_{\substack{i_1 + \cdots + i_k = n \\ i_j > 0}} \binom{b}{i_1} \cdots \binom{b}{i_k}.$$

Incidentally, the backward difference functional

$$f(t) = 1 - e^{-at}$$

satisfying

$$\langle f(t) \mid p(x) \rangle p(0) - p(a) \qquad \text{and} \qquad f(t)p(x) = p(x) - p(x - a)$$

is associated to the sequence of rising factorial polynomials

$$\left(\frac{x}{a} \right)^{(n)} = \left(\frac{x}{a} \right) \left(\frac{x}{a} + 1 \right) \cdots \left(\frac{x}{a} + n - 1 \right).$$

All of the above results have analogs for the rising factorial sequence, and we shall not give them here except to mention that the role of the Stirling numbers of the first kind $s(n, k)$ is taken by their absolute values $|s(n, k)|$, which are known as the signless Stirling numbers of the first kind, that is,

$$x^{(n)} = \sum_{k=0}^{n} |s(n, k)| x^k.$$

1.3. The Exponential Polynomials

DEFINITIONS

Let us consider the delta functional

$$f(t) = \log(1 + t).$$

Since $\bar{f}(t) = e^t - 1$ is the forward difference, Theorem 3.4.6 implies that the associated sequence for $f(t)$ is inverse to the lower factorial sequence under umbral composition. Equation (4.1.3) gives us the sequence

$$\phi_n(x) = \sum_{k=0}^{n} S(n, k)x^k.$$

These are called the *exponential polynomials*.

GENERATING FUNCTION AND SHEFFER IDENTITY

The generating function for the exponential polynomials is

$$\sum_{k=0}^{\infty} \frac{\phi_k(x)}{k!} t^k = e^{x(e^t - 1)}$$

and the binomial identity is

$$\phi_n(x + y) = \sum_{k=0}^{n} \binom{n}{k} \phi_k(x)\phi_{n-k}(y).$$

RECURRENCE RELATIONS AND OPERATIONAL FORMULAS

The recurrence formula is

$$\phi_{n+1}(x) = x[f'(t)]^{-1}\phi_n(x) = x(1 + t)\phi_n(x) = x(\phi_n(x) + \phi_n'(x)).$$

From this we also get the operational formula

$$\phi_n(x) = [x(1 + t)]^n 1.$$

But

$$e^{-x}(xt)e^x x^n = e^{-x}x(e^x x^n + ne^x x^{n-1}) = x(1 + t)x^n,$$

and so, as operators,

$$x(1 + t) = e^{-x}(xt)e^x.$$

This gives the operational formula

$$\phi_n(x) = e^{-x}(xt)^n e^x.$$

The second form of the recurrence formula is

$$\phi_{n+1}(x) = x\lambda_f[\bar{f}(t)]'x^n$$

$$= x\lambda_f e^t x^n$$

$$= x\lambda_f(x + 1)^n$$

$$= x \sum_{k=0}^{n} \binom{n}{k} \phi_k(x). \qquad (4.1.5)$$

From Theorem 2.4.5 we obtain

$$n\phi_{n-1}(x) = \log(1 + t)\phi_n(x)$$

$$= \sum_{k=1}^{\infty} \frac{(-1)^{k+1}}{k} t^k \phi_n(x)$$

$$= \sum_{k=1}^{n} \frac{(-1)^{k+1}}{k} \phi_n^{(k)}(x).$$

EXPANSIONS

The expansion theorem is

$$g(t) = \sum_{k=0}^{\infty} \frac{\langle g(t) \mid \phi_k(x) \rangle}{k!} [\log(1 + t)]^k.$$

For $g(t) = e^{yt}$ we get a formula for the exponential function in terms of the logarithm,

$$e^{yt} = \sum_{k=0}^{\infty} \frac{\phi_k(y)}{k!} [\log(1 + t)]^k.$$

Since $f'(\bar{f}(t)) = 1/(1 + (e^t - 1)) = e^{-t}$, Theorem 2.4.8 gives the expression

$$x\phi_n(x) = \sum_{k=1}^{n+1} \binom{n}{k-1} (-1)^{n-k+1} \phi_k(x).$$

Theorem 2.4.9 is

$$\phi_n'(x) = \sum_{k=0}^{n-1} \binom{n}{k} \langle e^t - 1 \mid x^{n-k} \rangle \phi_k(x) = \sum_{k=0}^{n-1} \binom{n}{k} \phi_k(x).$$

MISCELLANEOUS RESULTS

If we set $p_n(x) = (x)_n$, then $\phi_n(\mathbf{p}(x)) = x^n$ is just (4.1.3). But we also have

$$p_n(\boldsymbol{\phi}(x)) = x^n.$$

Writing

$$x^n = e^{-x} \sum_{k=0}^{\infty} \frac{(k)_n}{k!} x^k$$

(we are extending ourselves here a bit in writing this, but it can easily be justified), we have

$$p_n(\boldsymbol{\phi}(x)) = e^{-x} \sum_{k=0}^{\infty} \frac{p_n(k)}{k!} x^k.$$

Thus for any polynomial $p(x)$,

$$p(\phi(x)) = e^{-x} \sum_{k=0}^{\infty} \frac{p(k)}{k!} x^k.$$

If we take $p(x) = x^n$, then

$$\phi_n(x) = e^{-x} \sum_{k=0}^{\infty} \frac{k^n}{k!} x^k,$$

which is known as Dobinski's formula.

Let us turn again to the Stirling numbers. When the binomial identity for $\phi_n(x)$ is written out and the coefficients of like powers of x and y are compared, one obtains

$$\binom{i+j}{i} S(n, i+j) = \sum_{k=0}^{n} \binom{n}{k} S(k, i) S(n-k, j).$$

Since

$$\langle t^j \mid \phi_k(x) \rangle = j!\, S(k, j),$$

applying t^j to the recurrence (4.1.5) gives

$$j!\, S(n+1, j) = \left\langle t^j \,\middle|\, x \sum_{k=0}^{n} \binom{n}{k} \phi_k(x) \right\rangle$$

$$= \left\langle jt^{j-1} \,\middle|\, \sum_{k=0}^{n} \binom{n}{k} \phi_k(x) \right\rangle$$

$$= \sum_{k=0}^{n} \binom{n}{k} j!\, S(k, j-1)$$

or

$$S(n+1, j) = \sum_{k=0}^{n} \binom{n}{k} S(k, j-1).$$

Applying t^j to the expansion of $x\phi_n(x)$ gives, after dividing by $j!$,

$$S(n, j-1) = \sum_{k=1}^{n+1} \binom{n}{k-1} (-1)^{n-k+1} S(k, j).$$

The umbral composition $\phi_n(\mathbf{p}(x)) = x^n$, where $p_n(x) = (x)_n$, is equivalent to

$$x^n = \sum_{k=0}^{n} S(n, k)(x)_k$$

$$= \sum_{k=0}^{n} S(n, k) \sum_{j=0}^{k} s(k, j) x^j$$

$$= \sum_{j=0}^{n} \sum_{k=j}^{n} S(n, k) s(k, j) x^j,$$

and comparing coefficients of x^j gives

$$\delta_{n,j} = \sum_{k=j}^{n} S(n,k)s(k,j).$$

If we let $\lambda:x^n \to (x)_n$ be the umbral operator for $(x)_n$, then

$$(x)_n = \lambda^{-1} \circ \lambda(x)_n$$

$$= \sum_{k=0}^{n} s(n,k)\lambda^{-1} \circ \lambda x^k$$

$$= \sum_{k=0}^{n} s(n,k)\lambda^{-1}(x)_k$$

$$= \sum_{k=j}^{n} s(n,k) \sum_{j=0}^{k} s(k,j)\phi_j(x)$$

$$= \sum_{k=0}^{n} s(n,k) \sum_{j=0}^{k} s(k,j) \sum_{i=0}^{j} S(j,i)x^i$$

$$= \sum_{i=0}^{n} \sum_{k=i}^{n} \sum_{j=0}^{k} s(n,k)s(k,j)S(j,i)x^i,$$

and so

$$s(n,i) = \sum_{k=i}^{n} \sum_{j=0}^{k} s(n,k)s(k,j)S(j,i).$$

Many more properties of the Stirling numbers can be obtained by similar umbral methods.

1.4. The Gould Polynomials and the Central Factorial Polynomials

DEFINITIONS

We generalize the forward difference functional by setting

$$f(t) = e^{at}(e^{bt} - 1),$$

where $b \neq 0$.

Using Corollary 3.8.2 to relate the associated sequence for $f(t)$ to the lower factorial sequence, we have

$$q_n(x) = xe^{-ant}x^{-1}\left(\frac{x}{b}\right)_n = \frac{x}{x-an}\left(\frac{x-an}{b}\right)_n.$$

These are the *Gould polynomials*, which we denote by $G_n(x; a, b)$.

In case $a = -b/2$ we get the central difference series

$$\delta_b(t) = e^{bt/2} - e^{-bt/2},$$

which satisfies

$$\langle \delta_b(t) \mid p(x) \rangle = p(b/2) - p(-b/2)$$

and

$$\delta_b(t) p(x) = p(x + \tfrac{1}{2}b) - p(x - \tfrac{1}{2}b).$$

This series plays an important role in interpolation theory (Steffensen [2]).

The associated sequence for $\delta_1(t)$ is formed by the central factorial polynomials

$$x^{[n]} = G_n(x; -1/2, 1) = x(x + \tfrac{1}{2}n - 1)_{n-1}.$$

GENERATING FUNCTION AND SHEFFER IDENTITY

The generating function for the Gould polynomials is

$$\sum_{k=0}^{\infty} \frac{G_k(x; a, b)}{k!} t^k = e^{x \bar{f}(t)}.$$

By differentiating with respect to x and setting $x = 0$ we get

$$\bar{f}(t) = \sum_{k=1}^{\infty} \frac{G_k'(0; a, b)}{k!} t^k,$$

where $f(t) = e^{at}(e^{bt} - 1)$. Let us compute $G_k'(0; a, b)$ by umbral techniques,

$$G_k'(0; a, b) = \langle t \mid G_k(x; a, b) \rangle$$

$$= \left\langle t \;\middle|\; \frac{x}{x - ak} \left(\frac{x - ak}{b} \right)_k \right\rangle$$

$$= \left\langle t^0 \;\middle|\; \frac{1}{x - ak} \left(\frac{x - ak}{b} \right)_k \right\rangle$$

$$= \left\langle e^{-akt} \;\middle|\; \frac{1}{x} \left(\frac{x}{b} \right)_k \right\rangle$$

$$= \frac{1}{b} \left\langle e^{-akt} \;\middle|\; \left(\frac{x}{b} - 1 \right)_{k-1} \right\rangle$$

$$= \frac{1}{b} \left\langle e^{-(b + ak)t} \;\middle|\; \left(\frac{x}{b} \right)_{k-1} \right\rangle$$

$$= \frac{1}{b} \langle e^{-(b + ak)t} \mid \lambda x^{k-1} \rangle$$

(equation continues)

$$= \frac{1}{b} \langle \lambda * e^{-(b+ak)t} \mid x^{k-1} \rangle$$

$$= \frac{1}{b} \langle e^{-(b+ak)b^{-1}\log(1+t)} \mid x^{k-1} \rangle$$

$$= \frac{1}{b} \langle (1+t)^{-(b+ak)/b} \mid x^{k-1} \rangle$$

$$= \frac{1}{b} \sum_{j=0}^{\infty} \binom{-(b+ak)/b}{j} \langle t^j \mid x^{k-1} \rangle$$

$$= \frac{1}{b} \binom{-(b+ak)/b}{k-1} (k-1)!.$$

Thus we obtain

$$\bar{f}(t) = \sum_{k=1}^{\infty} \frac{1}{b} \binom{-(b+ak)/b}{k-1} \frac{t^k}{k}. \tag{4.1.6}$$

The binomial identity is a generalized Vandermonde convolution,

$$\frac{x+y}{x+y-an} \binom{(x+y-an)/b}{n}$$

$$= \sum_{k=0}^{n} \frac{x}{x-ak} \frac{y}{y-a(n-k)} \binom{(x-ak)/b}{k} \binom{(y-a(n-k))/a}{n-k}.$$

Such convolutions play an important role in combinatorics (Riordan [2]).

RECURRENCE RELATIONS AND OPERATIONAL FORMULAS

Equation (4.1.6) can be used in connection with the recurrence formula to give the recurrence

$$G_{n+1}(x; a, b) = x\lambda_f \bar{f}(t)'x^n$$

$$= x\lambda_f \sum_{k=1}^{\infty} \frac{1}{b} \binom{-(b+ak)/b}{k-1} t^{k-1} x^n$$

$$= x \sum_{k=1}^{n+1} \frac{1}{b} \binom{-(b+ak)/b}{k-1} (n)_{k-1} G_{n-k+1}(x; a, b).$$

For the central factorial polynomials, the first form of the recurrence formula is

$$x^{[n+1]} = x[f'(t)]^{-1} x^{[n]}$$

$$= 2x[e^{t/2} + e^{-t/2}]^{-1} x^{[n]}$$

$$= x\mu(t)^{-1} x^{[n]}, \tag{4.1.7}$$

where $\mu(t)$ is the averaging operator

$$\mu(t)p(x) = \frac{p(x + 1/2) + p(x - 1/2)}{2}.$$

Theorem 2.4.5 is

$$(x + 1/2)^{[n]} - (x - 1/2)^{[n]} = nx^{[n-1]}.$$

EXPANSIONS

The expansion theorem can be used to derive various classical interpolation formulas, such as those of Gauss, Stirling, Bessel, and Everett. We shall give only a brief sampling, referring the reader to Steffensen [2, Article 18] for more details.

The expansion theorem gives

$$e^{yt} = \sum_{k=0}^{\infty} \frac{y^{[k]}}{k!} \delta_1(t)^k,$$

and so

$$\frac{1}{2}(e^{yt} + e^{-yt}) = \sum_{k=0}^{\infty} \frac{1}{k!} \frac{y^{[k]} + (-y)^{[k]}}{2} \delta_1(t)^k = \sum_{k=0}^{\infty} \frac{y^{[2k]}}{(2k)!} \delta_1(t)^{2k}. \quad (4.1.8)$$

We wish to differentiate this with respect to $\delta_1(t)$. Using the chain rule (Theorem 3.6.3), we have

$$\partial_{\delta_1(t)}(e^{yt} + e^{-yt}) = \partial_{\delta_1(t)} t \, \partial_t(e^{yt} + e^{-yt})$$

$$= [\partial_t \delta_1(t)]^{-1} \partial_t(e^{yt} + e^{-yt})$$

$$= \mu(t)^{-1} y(e^{yt} - e^{-yt}),$$

and so

$$\frac{1}{2}\mu(t)^{-1} y(e^{yt} - e^{-yt}) = \sum_{k=1}^{\infty} \frac{y^{[2k]}}{(2k-1)!} \delta_1(t)^{2k-1}$$

$$= \sum_{k=0}^{\infty} \frac{y^{[2k+2]}}{(2k+1)!} \delta_1(t)^{2k+1}.$$

Thus we have

$$\frac{1}{2}(e^{yt} - e^{-yt}) = \sum_{k=0}^{\infty} \frac{y^{[2k+2]}}{(2k+1)!} \mu(t) \delta_1(t)^{2k+1}.$$

Adding this to (4.1.8) gives Stirling's formula

$$e^{yt} = \sum_{k=0}^{\infty} \left[\frac{y^{[2k]}}{(2k)!} \delta_1(t)^{2k} + \frac{y^{[2k+2]}/y}{(2k+1)!} \mu(t) \delta_1(t)^{2k+1} \right].$$

As another example, the expansion theorem gives

$$\mu(t)^{-1}f(t) = \sum_{k=0}^{\infty} \frac{\langle \mu(t)^{-1}f(t) \mid x^{[k]} \rangle}{k!} \delta_1(t)^k$$

for any series $f(t)$, and so from (4.1.7) we deduce that

$$f(t) = \sum_{k=0}^{\infty} \frac{\langle f(t) \mid x^{[k+1]}/x \rangle}{k!} \mu(t)\delta_1(t)^k.$$

Setting $f(t) = e^t$, we obtain a formula of Steffensen [1, p. 362]

$$e^{yt} = \sum_{k=0}^{\infty} \frac{y^{[k+1]}/y}{k!} \mu(t)\delta_1(t)^k.$$

MISCELLANEOUS RESULTS

Riordan [2, p. 212] gives a discussion of the numbers $t(n,k)$ and $T(n,k)$ defined by

$$x^{[n]} = \sum_{k=0}^{n} t(n,k)x^k$$

and

$$x^n = \sum_{k=0}^{n} T(n,k)x^{[k]}.$$

Note the analogy with the Stirling numbers. We leave it to the interested reader to obtain results analogous to those obtained earlier for the Stirling numbers.

We conclude this example with an amusing application of Proposition 2.1.3. If $p_n(x)$ is an associated sequence, we have

$$\langle [e^{at}(e^{bt} - 1)]^k \mid p_n(x) \rangle = \langle (e^{bt} - 1)^k \mid p_n(x + ak) \rangle$$

$$= \sum_{j=0}^{k} \binom{k}{j}(-1)^{k-j} p_n(bj + ak).$$

On the other hand, by Proposition 2.1.3,

$$\langle [e^{at}(e^{bt} - 1)]^k \mid p_n(x) \rangle$$

$$= \sum_{i_1 + \cdots + i_k = n} \binom{n}{i_1, \ldots, i_k} \langle e^{at}(e^{bt} - 1) \mid p_{i_1}(x) \rangle \cdots \langle e^{at}(e^{bt} - 1) \mid p_{i_k}(x) \rangle$$

$$= \sum_{i_1 + \cdots + i_k = n} \binom{n}{i_1, \ldots, i_k} [p_{i_1}(a + b) - p_{i_1}(a)] \cdots [p_{i_k}(a + b) - p_{i_k}(a)],$$

and so

$$\sum_{j=0}^{k} \binom{k}{j}(-1)^{k-j}p_n(bj + ak)$$

$$= \sum_{i_1 + \cdots + i_k = n} \binom{n}{i_1, \ldots, i_k}[p_{i_1}(a + b) - p_{i_1}(a)] \cdots [p_{i_k}(a + b) - p_{i_k}(a)].$$

Taking $p_n(x) = x^n$, $a = 1$ and $a + b = -1$, we get

$$\sum_{j=0}^{k} \binom{k}{j}(-1)^{k-j}(k - 2j)^n$$

$$= \sum_{i_1 + \cdots + i_k = n} \binom{n}{i_1, \ldots, i_k}[(-1)^{i_1} - 1] \cdots [(-1)^{i_k} - 1]$$

$$= (-2)^k \sum_{\substack{i_1 + \cdots + i_k = n \\ i_j \text{ odd}}} \binom{n}{i_1, \ldots, i_k}.$$

Except for the factor $(-2)^k$, the right side counts the number of ways of distributing n balls into k boxes, subject to the condition that each box receive an odd number of balls!

1.5. The Abel Polynomials

DEFINITIONS

The associated sequence for the Abel functional

$$f(t) = te^{at} \qquad (a \neq 0)$$

can be determined from the transfer formula to be

$$A_n(x; a) = x\left(\frac{f(t)}{t}\right)^{-n} x^{n-1}$$

$$= xe^{-ant}x^{n-1}$$

$$= x(x - an)^{n-1}.$$

These are the Abel polynomials.

GENERATING FUNCTION AND SHEFFER IDENTITY

The generating function for the Abel polynomials is

$$\sum_{k=0}^{\infty} \frac{x(x - ak)^{k-1}}{k!} t^k = e^{x\bar{f}(t)},$$

where $f(t) = te^{at}$. Differentiating with respect to x and setting $x = 0$ give

$$\bar{f}(t) = \sum_{k=1}^{\infty} \frac{(-ak)^{k-1}}{k!} t^k.$$

The binomial identity in this case is

$$(x + y)(x + y - an)^{n-1} = \sum_{k=0}^{n} \binom{n}{k} x y(x - ak)^{k-1}(y - a(n - k))^{n-k-1}.$$

There are a host of identities of this form, and Riordan [2, p. 18] gives a thorough account. For example, one of Abel's generalizations of the binomial formula is [Riordan 2, p. 18, Eq. (13)]

$$x^{-1}(x + y - na)^n = \sum_{k=0}^{n} \binom{n}{k}(x - ak)^{k-1}(y - a(n - k))^{n-k}. \qquad (4.1.9)$$

Since

$$te^{at}(x - an)^n = ne^{at}(x - an)^{n-1} = n(x - a(n - 1))^{n-1},$$

we see that the sequence $s_n(x) = (x - an)^n$ is nothing but a Sheffer sequence for the Abel functional $f(t) = te^{at}$ and that (4.1.9) is nothing but the Sheffer identity for $s_n(x)$.

RECURRENCE RELATIONS AND OPERATIONAL FORMULAS

The recurrence formula for Abel polynomials is

$$\begin{aligned}
A_{n+1}(x; a) &= x[f'(t)]^{-1} A_n(x; a) \\
&= x(1 + at)^{-1} e^{-at} A_n(x; a) \\
&= x(1 + at)^{-1} A_n(x - a; a) \\
&= x \sum_{k=0}^{n} a^k A_n^{(k)}(x - a; a),
\end{aligned}$$

and the second form gives

$$\begin{aligned}
A_{n+1}(x; a) &= x\lambda_f[\bar{f}(t)]' x^n \\
&= x\lambda_f \sum_{k=1}^{\infty} \frac{(-ak)^{k-1}}{(k - 1)!} t^{k-1} x^n \\
&= x \sum_{k=1}^{n+1} \binom{n}{k - 1}(-ak)^{k-1} A_{n-k+1}(x; a).
\end{aligned}$$

Theorem 2.4.5 is the simple equation

$$A_n'(x + a; a) = nA_{n-1}(x; a).$$

EXPANSIONS

The expansion theorem is

$$g(t) = \sum_{k=0}^{\infty} \frac{\langle g(t) \mid x(x - ak)^{k-1} \rangle}{k!} t^k e^{akt}.$$

Taking $g(t) = e^{yt}$, we get

$$e^{yt} = \sum_{k=0}^{\infty} \frac{y(y - ak)^{k-1}}{k!} t^k e^{akt},$$

and for any polynomial

$$p(x + y) = \sum_{k \geq 0} \frac{y(y - ak)^{k-1}}{k!} p^{(k)}(x + ak).$$

Taking $g(t) = t^n$ in the expansion theorem and observing that for $k \geq n \geq 1$

$$\langle t^n \mid x(x - ak)^{k-1} \rangle = \langle nt^{n-1} \mid (x - ak)^{k-1} \rangle$$

$$= n \sum_{j=0}^{k-1} \binom{k-1}{j} (-ak)^{k-1-j} \langle t^{n-1} \mid x^j \rangle$$

$$= n(k - 1)_{n-1}(-ak)^{k-n},$$

we have

$$t^n = \sum_{k=n}^{\infty} \frac{n(k-1)_n}{k!} (-ak)^{k-n} t^k e^{akt}.$$

Applying this to a polynomial $p(x)$ gives a formula for $p^{(n)}(0)$ in terms of $p^{(k)}(ak)$,

$$p^{(n)}(0) = \sum_{k \geq n} \frac{n(k-1)_{n-1}}{k!} (-ak)^{k-n} p^{(k)}(ak).$$

The polynomial expansion theorem is

$$p(x) = \sum_{k \geq 0} \frac{\langle t^k e^{akt} \mid p(x) \rangle}{k!} x(x - ak)^{k-1}$$

or

$$p(x) = \sum_{k \geq 0} \frac{p^{(k)}(ak)}{k!} x(x - ak)^{k-1}.$$

Taking, for instance, $p(x) = x^n$ gives

$$x^n = \sum_{k=1}^{n} \binom{n}{k} (ak)^{n-k} x(x - ak)^{k-1}.$$

This shows that the inverse, under umbral composition, of the Abel sequence $A_n(x;a)$ is the sequence

$$q_n(x) = \sum_{k=1}^{n} \binom{n}{k}(ak)^{n-k}x^k.$$

This sequence is associated to $\bar{f}(t) = \sum_{k=1}^{\infty}(-ak)^{k-1}t^k/k!$.

Theorem 2.4.8 is

$$xA_n(x;a) = \sum_{k=1}^{n+1}\binom{n}{k-1}\langle(1+at)e^{at}\,|\,A_{n-k+1}(x;a)\rangle A_k(x;a)$$

$$= \sum_{k=1}^{n+1}\binom{n}{k-1}(A_{n-k+1}(a;a) + a\,\delta_{n,k})A_k(x;a)$$

$$= a\sum_{k=1}^{n+1}\binom{n}{k-1}([-a(n-k)]^{n-k} + \delta_{n,k})A_k(x;a).$$

Theorem 2.4.9 is

$$A_n'(x;a) = \sum_{k=0}^{n-1}\binom{n}{k}\langle t\,|\,A_{n-k}(x;a)\rangle A_k(x;a)$$

$$= \sum_{k=0}^{n-1}\binom{n}{k}[-a(n-k)]^{n-k-1}A_k(x;a).$$

1.6. The Mittag-Leffler Polynomials

DEFINITIONS

We next consider the delta series

$$f(t) = \frac{e^t - 1}{e^t + 1}$$

and note that

$$f'(t) = \frac{2e^t}{(e^t + 1)^2},$$

$$\bar{f}(t) = \log\left(\frac{1+t}{1-t}\right).$$

Let us denote the associated sequence for $f(t)$ by $M_n(x)$. We call the $M_n(x)$ Mittag-Leffler polynomials (Bateman [1], Erdelyi [1, Vol. 3, p. 248]). It should be noted that both the aforementioned sources define the Mittag-Leffler polynomials as $M_n(x)/n!$.

We obtain a nice expression for the Mittag-Leffler polynomials by using Theorem 3.5.6,

$$M_n(y) = \langle e^{yt} \mid M_n(x) \rangle$$

$$= \left\langle \exp\left(y \log\left(\frac{1+t}{1-t}\right) \right) \middle| x^n \right\rangle$$

$$= \left\langle \left(\frac{1+t}{1-t}\right)^y \middle| x^n \right\rangle$$

$$= \left\langle \left(1 + \frac{2t}{1-t}\right)^y \middle| x^n \right\rangle$$

$$= \sum_{k=0}^{n} \binom{y}{k} 2^k \left\langle \frac{t^k}{(1-t)^k} \middle| x^n \right\rangle$$

$$= \sum_{k=0}^{n} \binom{y}{k} 2^k (n)_k \langle (1-t)^{-k} \mid x^{n-k} \rangle$$

$$= \sum_{k=0}^{n} \binom{y}{k} 2^k (n)_k (n-1)_{n-k},$$

and so

$$M_n(x) = \sum_{k=0}^{n} \binom{n}{k}\binom{n-1}{n-k} 2^k (x)_k.$$

GENERATING FUNCTION AND SHEFFER IDENTITY

The generating function for the Mittag-Leffler polynomials is

$$\sum_{k=0}^{\infty} \frac{M_k(x)}{k!} t^k = \left(\frac{1+t}{1-t}\right)^x.$$

These polynomials were used by Mittag-Leffler in the study of the integrals and invariants of linear homogeneous differential equations, wherein he used the conformal mapping

$$w = ((1+z)/(1-z))^x$$

for x constant. They were also used by Pidduck (see Bateman [1]) to study the propagation of a disturbance in a fluid acted on by gravity.

The binomial identity is

$$M_n(x+y) = \sum_{k=0}^{n} \binom{n}{k} M_k(x) M_{n-k}(y).$$

RECURRENCE RELATIONS AND OPERATIONAL FORMULAS

Theorem 2.4.5 is

$$\frac{e^t - 1}{e^t + 1} M_n(x) = nM_{n-1}(x),$$

which is equivalent to

$$(e^t - 1)M_n(x) = n(e^t + 1)M_{n-1}(x)$$

or

$$M_n(x + 1) - M_n(x) = n[M_{n-1}(x + 1) + M_{n-1}(x)].$$

The recurrence formula is

$$
\begin{aligned}
M_{n+1}(x) &= x[f'(t)]^{-1}M_n(x)\\
&= \tfrac{1}{2}xe^{-t}(e^t + 1)^2 M_n(x)\\
&= \tfrac{1}{2}x(e^t + 2 + e^{-t})M_n(x)\\
&= \tfrac{1}{2}x[M_n(x + 1) + 2M_n(x) + M_n(x - 1)],
\end{aligned}
$$

and the second form gives the recurrence

$$
\begin{aligned}
M_{n+1}(x) &= x\lambda_f[\bar{f}(t)]'x^n\\
&= x\lambda_f(2/(1 - t^2))x^n\\
&= 2x\lambda_f \sum_{k=0}^{\infty} t^{2k}x^n\\
&= 2x \sum_{k\geq 0} (n)_{2k}M_{n-2k}.
\end{aligned}
$$

EXPANSIONS

The expansion theorem for the Mittag-Leffler polynomials is

$$g(t) = \sum_{k=0}^{\infty} \frac{\langle g(t)\,|\, M_k(x)\rangle}{k!}\left(\frac{e^t - 1}{e^t + 1}\right)^k,$$

and since $\langle g(t)\,|\, M_k(x)\rangle = \langle g(\bar{f}(t))\,|\, x^k\rangle$,

$$g(t) = \sum_{k=0}^{\infty} \frac{1}{k!}\left\langle g\left(\log\left(\frac{1 + t}{1 - t}\right)\right)\,\Big|\, x^k\right\rangle\left(\frac{e^t - 1}{e^t + 1}\right)^k.$$

For $g(t) = e^{yt}$ we get

$$e^{yt} = \sum_{k=0}^{\infty}\left[\sum_{j=0}^{k}\binom{k}{j}\binom{k - 1}{k - j}2^j(y)_j\right]\left(\frac{e^t - 1}{e^t + 1}\right)^k.$$

Since $f'(\bar{f}(t)) = (1 - t^2)/2$, Theorem 2.4.8 yields

$$xM_n(x) = \sum_{k=1}^{n+1} \binom{n}{k-1} \left\langle \frac{1-t^2}{2} \middle| x^{n-k+1} \right\rangle M_k(x)$$

$$= \frac{1}{2}[M_{n+1}(x) - n(n-1)M_{n-1}(x)].$$

Theorem 2.4.9 gives

$$M'_n(x) = \sum_{k=0}^{n-1} \binom{n}{k} \langle t \,|\, M_{n-k}(x) \rangle M_k(x),$$

and since

$$\langle t \,|\, M_{n-k}(x) \rangle = \sum_{j=0}^{n-k} \binom{n-k}{j}\binom{n-k-1}{n-k-j} 2^j \langle t \,|\, (x)_j \rangle$$

$$= -\sum_{j=1}^{n-k} \binom{n-k}{j}\binom{n-k-1}{n-k-j}(-2)^j(j-1)!,$$

we get

$$M'_n(x) = -\sum_{k=0}^{n-1}\left[\sum_{j=1}^{n-k}\binom{n}{k}\binom{n-k}{j}\binom{n-k-1}{n-k-j}(-2)^j(j-1)!\right]M_k(x).$$

1.7. The Bessel Polynomials

DEFINITIONS

Krall and Frink [1] made a study of the polynomials

$$y_n(x) = \sum_{k=0}^{n} \frac{(n+k)!}{(n-k)!\,k!}\left(\frac{x}{2}\right)^k,$$

which are called Bessel polynomials in view of their connection with Bessel functions. The $y_n(x)$ satisfy the differential equation

$$x^2 y'' + (2x + 2)y' + n(n+1)y = 0,$$

and their derivatives form an orthogonal sequence of polynomials. Krall and Frink also discuss applications of these polynomials to spherical waves.

Carlitz [2] defined a related set of polynomials

$$p_n(x) = x^n y_{n-1}(1/x)$$

and gave many of their properties. Now, by the transfer formula, the associated sequence for

$$f(t) = t - t^2/2$$

is

$$x\left(\frac{f(t)}{t}\right)^{-n} x^{n-1} = x\left(1 - \frac{t}{2}\right)^{-n} x^{n-1}$$

$$= x \sum_{k=0}^{n-1} \binom{-n}{k} \left(-\frac{1}{2}\right)^k (n-1)_k x^{n-1-k}$$

$$= \sum_{k=0}^{n-1} \frac{(n-1+k)!}{(n-1-k)!\,k!} \left(\frac{1}{2}\right)^k x^{n-k}$$

$$= x^n y_{n-1}\left(\frac{1}{x}\right).$$

Thus

$$p_n(x) = x^n y_{n-1}\left(\frac{1}{x}\right)$$

$$= \sum_{k=1}^{n} \frac{(2n-k-1)!}{(k-1)!\,(n-k)!} \left(\frac{1}{2}\right)^{n-k} x^k$$

is associated to $f(t) = t - t^2/2$.

GENERATING FUNCTION AND SHEFFER IDENTITY

The generating function for $p_n(x)$ is

$$\sum_{k=0}^{\infty} \frac{p_k(x)}{k!} t^k = e^{x[1 - (1 - 2t)^{1/2}]}$$

and the binomial identity is

$$p_n(x + y) = \sum_{k=0}^{n} \binom{n}{k} p_k(x) p_{n-k}(y),$$

which translates into (provided we set $y_{-1}(x) = 1$)

$$(x + y)^n y_{n-1}\left(\frac{1}{x+y}\right) = \sum_{k=0}^{n} \binom{n}{k} x^k y^{n-k} y_{k-1}\left(\frac{1}{x}\right) y_{n-k-1}\left(\frac{1}{y}\right).$$

RECURRENCE RELATIONS AND OPERATIONAL FORMULAS

According to Theorem 2.4.5, we have

$$\left(t - (t^2/2)\right) p_n(x) = n p_{n-1}(x)$$

or

$$p_n''(x) - 2p_n'(x) + 2n p_{n-1}(x) = 0.$$

The recurrence formula gives

$$p_{n+1}(x) = x[f'(t)]^{-1}p_n(x)$$
$$= x(1 - t)^{-1}p_n(x)$$
$$= x \sum_{k=0}^{n} p_n^{(k)}(x)$$

as well as

$$p_n(x) = (1 - t)\frac{p_{n+1}(x)}{x},$$

or, by performing the indicated derivative and simplifying,

$$(x + 1)p_{n+1}(x) - xp'_{n+1}(x) - x^2p_n(x) = 0.$$

From the second form of the recurrence formula we obtain

$$p_{n+1}(x) = x\lambda_f[\bar{f}(t)]'x^n$$
$$= x\lambda_f(1 - 2t)^{-1/2}x^n$$
$$= x\lambda_f \sum_{k=0}^{\infty} \binom{-1/2}{k}(-2)^k(n)_k x^{n-k}$$
$$= x \sum_{k=0}^{n} \binom{n}{k}[1 \cdot 3 \cdots (2k - 3)(2k - 1)]x^{n-k}.$$

EXPANSIONS

The polynomial expansion theorem is

$$p(x) = \sum_{k \geq 0} \frac{\langle (t - t^2/2)^k \mid p(x) \rangle}{k!} p_k(x).$$

But

$$\langle (t - t^2/2)^k \mid p(x) \rangle = \langle (1 - t/2)^k \mid p^{(k)}(x) \rangle$$
$$= \sum_{j=0}^{k} \binom{k}{j}\left(-\frac{1}{2}\right)^j \langle t^j \mid p^{(k)}(x) \rangle$$
$$= \sum_{j=0}^{k} \binom{k}{j}\left(-\frac{1}{2}\right)^j p^{(k+j)}(0),$$

and so

$$p(x) = \sum_{k \geq 0}\left[\sum_{j=0}^{k} \frac{1}{k!}\binom{k}{j}\left(-\frac{1}{2}\right)^j p^{(k+j)}(0)\right]p_k(x).$$

Taking $p(x) = x^n$ gives

$$x^n = \sum_{k \geq n/2} \binom{k}{n-k} \frac{n!}{k!} \left(-\frac{1}{2}\right)^{n-k} p_k(x).$$

Theorem 2.4.8 gives

$$xp_n(x) = \sum_{k=1}^{n+1} \binom{n}{k-1} \langle (1-t) \mid p_{n-k+1}(x) \rangle p_k(x)$$

$$= \sum_{k=1}^{n+1} \binom{n}{k-1} \left[\delta_{k,n+1} - \frac{(2n-2k)!}{(n-k)!} \left(\frac{1}{2}\right)^{n-k} \right] p_k(x)$$

$$= p_{n+1}(x) - \sum_{k=1}^{n} \binom{n}{k-1} \frac{(2n-2k)!}{(n-k)!} \left(\frac{1}{2}\right)^{n-k} p_k(x).$$

Theorem 2.4.9 yields

$$p_n'(x) = \sum_{k=0}^{n-1} \binom{n}{k} \langle t \mid p_{n-k}(x) \rangle p_k(x)$$

$$= \sum_{k=0}^{n-1} \binom{n}{k} \frac{(2n-2k-2)!}{(n-k-1)!} \left(\frac{1}{2}\right)^{n-k-1} p_k(x).$$

Let us also expand $x^2 p_n(x)$ as

$$x^2 p_n(x) = \sum_{k=0}^{n+2} \frac{\langle (t-t^2/2)^k \mid x^2 p_n(x) \rangle}{k!} p_k(x).$$

Now

$$\langle (t - t^2/2)^k \mid x^2 p_n(x) \rangle$$

$$= \langle \partial_t^2 [(t - t^2/2)^k] \mid p_n(x) \rangle$$

$$= \langle k(k-1)(t - t^2/2)^{k-2}(1-t)^2 - k(t - t^2/2)^{k-1} \mid p_n(x) \rangle$$

$$= k(k-1)(n)_{k-2} \langle (1-t)^2 \mid p_{n-k+2}(x) \rangle - k! \, \delta_{k,n+1}$$

$$= k(k-1)(n)_{k-2} \langle (1 - [1 - (1-2t)^{1/2}])^2 \mid x^{n-k+2} \rangle - k! \, \delta_{k,n+1}$$

$$= k(k-1)(n)_{k-2} \langle 1 - 2t \mid x^{n-k+2} \rangle - k! \, \delta_{k,n+1}$$

$$= k(k-1)(n)_{k-2}(\delta_{k,n+2} - 2\delta_{k,n+1}) - k! \delta_{k,n+1}$$

$$= k! \, \delta_{k,n+2} - (2n+1)k! \, \delta_{k,n+1},$$

and so

$$x^2 p_n(x) = p_{n+2}(x) - (2n+1)p_{n+1}(x)$$

or

$$p_{n+1}(x) = (2n-1)p_n(x) + x^2 p_{n-1}(x).$$

1.8. The Bell Polynomials

DEFINITIONS

Let us consider a somewhat different type of example. Let K be a field of characteristic zero and let $x_1, x_2, \ldots$ be a sequence of independent variables. We take the field C to be K with the variables x_i adjoined,

$$C = K(x_1, x_2, \ldots).$$

The *generic delta series* is the series

$$g(t) = \sum_{k=1}^{\infty} \frac{x_k}{k!} t^k,$$

and

$$\langle g(t) \mid x^n \rangle = x_n$$

for all $n \geq 1$.

Let $b_n(x; x_1, x_2, \ldots)$ denote the associated sequence for $\bar{g}(t)$, the compositional inverse of the generic delta functional, and set

$$b_n(x; x_1, x_2, \ldots) = \sum_{k=0}^{n} B_{n,k}(x_1, x_2, \ldots) x^k.$$

The sequence $b_n(x; x_1, x_2, \ldots)$ is the *generic associated sequence*, so named since any associated sequence is but a special case of $b_n(x; x_1, x_2, \ldots)$ for an appropriate choice of the variables x_k. We shall see some examples in the subsection on miscellaneous results.

Results obtained for the sequence $b_n(x; x_1, x_2, \ldots)$ are therefore generic results that apply to all associated sequences. The same is true for the generic coefficients $B_{n,k}(x_1, x_2, \ldots)$.

The conjugate representation provides an explicit formula for the coefficients $B_{n,k}(x_1, x_2, \ldots)$,

$$B_{n,k}(x_1, x_2, \ldots) = \frac{1}{k!} \langle g(t)^k \mid x^n \rangle$$

$$= \frac{1}{k!} \sum_{i_1 + \cdots + i_k = n} \binom{n}{i_1, \ldots, i_k} \langle g(t) \mid x^{i_1} \rangle \cdots \langle g(t) \mid x^{i_k} \rangle$$

$$= \frac{1}{k!} \sum_{\substack{i_1 + \cdots + i_k = n \\ i_j > 0}} \binom{n}{i_1, \ldots, i_k} x_{i_1} \cdots x_{i_k}$$

$$= \sum_{\substack{j_1 + j_2 + \cdots = k \\ j_1 + 2j_2 + \cdots = n}} \frac{n!}{j_1! j_2! \cdots} \left(\frac{x_1}{1!} \right)^{j_1} \left(\frac{x_2}{2!} \right)^{j_2} \cdots.$$

The $B_{n,k}(x_1, x_2, \ldots)$ are known as the Bell polynomials (Comtet [1, p. 133]), and they have important combinatorial significance.

GENERATING FUNCTION AND SHEFFER IDENTITY

The generating function for $b_n(x; x_1, x_2, \ldots)$ is

$$\sum_{k=0}^{\infty} \frac{b_k(x; x_1, x_2, \ldots)}{k!} t^k = e^x \left(\sum_{k=1}^{\infty} \frac{x_k}{k!} t^k \right).$$

The binomial identity is

$$b_n(x + y; x_1, x_2, \ldots) = \sum_{k=0}^{n} \binom{n}{k} b_k(x; x_1, x_2, \ldots) b_{n-k}(y; x_1, x_2, \ldots),$$

which is equivalent to the following identity for the coefficients,

$$\binom{i+j}{i} B_{n,i+j}(x_1, x_2, \ldots) = \sum_{k=0}^{n} \binom{n}{k} B_{k,i}(x_1, x_2, \ldots) B_{n-k,j}(x_1, x_2, \ldots).$$

$$(4.1.10)$$

This generic result shows that a sequence

$$p_n(x) = \sum_{k=0}^{n} a_{n,k} x^k$$

is an associated sequence if and only if the coefficients $a_{n,k}$ satisfy

$$\binom{i+j}{i} a_{n,i+j} = \sum_{k=0}^{n} \binom{n}{k} a_{k,i} a_{n-k,j}.$$

Taking $i = 1$ and $j = l - 1$ in (4.1.10), we get the recurrence formula

$$l B_{n,l}(x_1, x_2, \ldots) = \sum_{k=1}^{n} \binom{n}{k} x_k B_{n-k, l-1}(x_1, x_2, \ldots).$$

RECURRENCE RELATIONS AND OPERATIONAL FORMULAS

The second form of the recurrence formula is

$$b_n(x; x_1, x_2, \ldots) = x \lambda_g [g(t)]' x^{n-1}$$

$$= x \lambda_g \sum_{j=1}^{\infty} \frac{x_j}{(j-1)!} t^{j-1} x^{n-1} = x \lambda_g \sum_{j=1}^{n} \binom{n-1}{j-1} x_j x^{n-j}$$

$$= x \sum_{j=1}^{n} \binom{n-1}{j-1} x_j b_{n-j}(x; x_1, x_2, \ldots).$$

Applying t^k to both sides gives the recurrence

$$B_{n,k}(x_1, x_2, \ldots) = \sum_{j=k-1}^{n-1} \binom{n-1}{j} x_{n-j} B_{j,k-1}(x_1, x_2, \ldots).$$

MISCELLANEOUS RESULTS

Other identities involving the Bell polynomials follow easily by umbral methods. For example, setting

$$\langle g_1(t) \mid x^n \rangle = x_{n+1}/(n+1) \tag{4.1.11}$$

for $n \geq 1$ and $\langle g_1(t) \mid x^0 \rangle = 0$, we have

$$g(t) = t(x_1 + g_1(t)).$$

Hence

$$g(t)^k = \sum_{j=0}^{k} \binom{k}{j} x_1^{k-j} g_1(t)^j t^k. \tag{4.1.12}$$

From (4.1.11) we see that $\langle g_1(t)^k \mid x^n \rangle = k! \, B_{n,k}(x_2/2, x_3/3, \ldots)$, and so applying (4.1.12) to x^n gives the identity

$$B_{n,k}(x_1, x_2, \ldots) = \sum_{j=0}^{k} \frac{n!}{(n-k)!(k-j)!} x_1^{k-j} B_{n-k,j}\left(\frac{x_2}{2}, \frac{x_3}{3}, \ldots\right).$$

If we adjoin additional variables $y_1, y_2, \ldots$ to C and set

$$\langle h(t) \mid x^n \rangle = x_n + y_n$$

for $n \geq 1$ and $\langle h(t) \mid x^0 \rangle = 0$, then the associated sequence for $\bar{h}(t)$ is $b_n(x; x_1 + y_1, x_2 + y_2, \ldots)$. Now

$$\langle h(t)^k \mid x^n \rangle = \langle [g(t) + (h(t) - g(t))]^k \mid x^n \rangle$$

$$= \sum_{j=0}^{k} \binom{k}{j} \langle g(t)^j \mid x^n \rangle \langle (h(t) - g(t))^{k-j} \mid x^n \rangle,$$

whence we obtain

$$B_{n,k}(x_1 + y_1, x_2 + y_2, \ldots) = \sum_{j=0}^{k} B_{n,j}(x_1, x_2, \ldots) B_{n,k-j}(y_1, y_2, \ldots).$$

Let us consider some special cases of the Bell polynomials. If we set

$$x_k = 1$$

for all $k \geq 1$, then the generic functional becomes

$$g(t) = \sum_{k=1}^{\infty} \frac{1}{k!} t^k = e^t - 1,$$

and so

$$b_n(x; 1, 1, \ldots) = \phi_n(x),$$

where $\phi_n(x)$ are the exponential polynomials. Thus

$$B_{n,k}(1, 1, \ldots) = S(n, k)$$

are the Stirling numbers of the second kind.

If we take

$$x_k = (-1)^{k-1}(k - 1)!$$

for all $k \geq 1$, then

$$g(t) = \sum_{k=1}^{\infty} (-1)^{k-1} \frac{(k - 1)!}{k!} t^k = \log(1 + t),$$

and so

$$g(t) = (x)_n.$$

Thus

$$B_{n,k}(0!, -1!, 2!, -3!, \ldots) = s(n, k)$$

are the Stirling numbers of the first kind.

If we take

$$x_k = ka^{k-1}$$

for all $k \geq 1$, then

$$g(t) = \sum_{k=1}^{\infty} \frac{a^{k-1}}{(k - 1)!} t^k = te^{at},$$

which is the Abel functional. As we saw in the subsection on expansions in Section 1.5, the associated sequence for $\bar{g}(t)$ is

$$b_n(x; 1, 2a, 3a^2, \ldots) = \sum_{k=1}^{n} \binom{n}{k} (ak)^{n-k} x^k.$$

Thus

$$B_{n,k}(1, 2a, 3a^2, \ldots) = \binom{n}{k} (ak)^{n-k}.$$

For $a = 1$ these numbers are known as the idempotent numbers (Comtet [1, p. 91]).

Finally, let us take

$$x_k = k!$$

for all $k \geq 1$. Then

$$g(t) = \sum_{k=1}^{\infty} t^k = \frac{t}{1-t} \quad \text{and} \quad \bar{g}(t) = \frac{t}{1+t}.$$

The transfer formula gives us the associated sequence for $\bar{g}(t)$,

$$b_n(x; 1!, 2!, 3!, \ldots) = x(1+t)^n x^{n-1}$$

$$= x \sum_{k=0}^{n-1} \binom{n}{k} (n-1)_k x^{n-1-k}$$

$$= \sum_{k=1}^{n} \binom{n-1}{k-1} \frac{n!}{k!} x^k.$$

Thus

$$B_{n,k}(1!, 2!, 3!, \ldots) = \binom{n-1}{k-1} \frac{n!}{k!}.$$

These are the Lah numbers (Comtet [1, p. 156]). We shall encounter them again in connection with the Laguerre polynomials.

2. APPELL SEQUENCES

For ease of reference we list the key results. Let $s_n(x)$ be Appell for $g(t)$.

Generating function

$$\sum_{k=0}^{\infty} \frac{s_k(x)}{k!} t^k = \frac{1}{g(t)} e^{xt}.$$

Appell identity

$$s_n(x+y) = \sum_{k=0}^{n} \binom{n}{k} s_k(y) x^{n-k}.$$

Theorem 2.5.5

$$s_n(x) = g(t)^{-1} x^n.$$

Theorem 2.5.6

$$t s_n(x) = n s_{n-1}(x).$$

Recurrence formula

$$s_{n+1}(x) = \left(x - \frac{g'(t)}{g(t)} \right) s_n(x).$$

Expansion theorem

$$h(t) = \sum_{k=0}^{\infty} \frac{\langle h(t) \,|\, s_k(x) \rangle}{k!} g(t) t^k.$$

Polynomial expansion theorem

$$p(x) = \sum_{k \geq 0} \frac{\langle g(t) \,|\, t^k p(x) \rangle}{k!} s_k(x).$$

Theorem 3.5.6

$$\langle h(t) \,|\, q(\mathbf{s}(x)) \rangle = \langle g(t)^{-1} h(t) \,|\, q(x) \rangle,$$

$$\langle h(t) \,|\, s_n(x) \rangle = \langle g(t)^{-1} h(t) \,|\, x^n \rangle.$$

Theorem 2.5.9

$$x s_n(x) = s_{n+1}(x) + \sum_{k=0}^{n} \binom{n}{k} \langle g'(t) \,|\, s_{n-k}(x) \rangle s_k(x)$$

or

$$x s_n(x) = s_{n+1}(x) + \sum_{k=0}^{n} \binom{n}{k} \langle g'(t)/g(t) \,|\, x^{n-k} \rangle s_k(x).$$

Conjugate representation

$$s_n(x) = \sum_{k=0}^{n} \binom{n}{k} \langle g(t)^{-1} \,|\, x^{n-k} \rangle x^k.$$

Multiplication theorem

$$s_n(\alpha x) = \alpha^n \frac{g(t)}{g(t/\alpha)} s_n(x) \qquad (\alpha \neq 0).$$

Proposition 2.1.11

$$\langle h(t) \,|\, p(ax) \rangle = \langle h(at) \,|\, p(x) \rangle.$$

Theorem 2.1.10

$$\langle h(t) \,|\, xp(x) \rangle = \langle \partial_t h(t) \,|\, p(x) \rangle.$$

2.1. The Hermite Polynomials

DEFINITIONS

The Hermite polynomials $H_n^{(v)}(x)$ of variance v form the Appell sequence for

$$g(t) = e^{vt^2/2}.$$

The case $v = 1$ is the most familiar, and we write $H_n^{(1)}(x) = H_n(x)$.

Theorem 2.5.5 easily gives an explicit expression for the Hermite polynomials.

$$H_n^{(v)}(x) = e^{-vt^2/2}x^n = \sum_{k=0}^{\infty} \left(\frac{-v}{2}\right)^k \frac{1}{k!} t^{2k} x^n$$

$$= \sum_{k=0}^{[n/2]} \left(\frac{-v}{2}\right)^k \frac{(n)_{2k}}{k!} x^{n-2k}.$$

From this we deduce that

$$H_n^{(v)}(x) = v^{n/2} H_n(x/v^{1/2}).$$

Incidentally, the operator $e^{vt^2/2}$ satisfies, for $v > 0$,

$$e^{vt^2/2}p(x) = \frac{1}{(2\pi v)^{1/2}} \int_{-\infty}^{\infty} e^{-u^2/2v} p(x + u)\, du$$

and has been called the Weierstrass operator.

GENERATING FUNCTION AND SHEFFER IDENTITY

The generating function for the Hermit polynomials is

$$\sum_{k=0}^{\infty} \frac{H_k^{(v)}(x)}{k!} t^k = e^{xt - vt^2/2}.$$

It should be noted that many sources, including Erdelyi [1, Vol. 2, p. 192], define the Hermite polynomials by means of the generating function

$$\sum_{k=0}^{\infty} \frac{u_k(x)}{k!} t^k = e^{2xt - vt^2}.$$

According to this definition, the sequence $k!\, u_k(x)$ is Sheffer, but it is not Appell, for the delta functional is $f(t) = t/2$. In any case, since

$$u_n(x) = 2^{n/2} H_n^{(v)}(2^{1/2}x),$$

the discrepancy in the definitions is only minor (even though it is a major nuisance).

The Appell identity in this case is

$$H_n^{(v)}(x + y) = \sum_{k=0}^{n} \binom{n}{k} H_n^{(v)}(y) x^{n-k},$$

and applying $e^{-\mu t^2/2}$ gives the appealing identity

$$H_n^{(v+\mu)}(x + y) = \sum_{k=0}^{n} \binom{n}{k} H_k^{(v)}(y) H_{n-k}^{(\mu)}(x). \tag{4.2.1}$$

Taking $\mu = -v$ gives

$$(x + y)^n = \sum_{k=0}^{n} \binom{n}{k} H_k^{(v)}(x) H_{n-k}^{(-v)}(y).$$

We shall say more about identities of the form (4.2.1) when we discuss cross sequences in Chapter 5.

RECURRENCE RELATIONS AND OPERATIONAL FORMULAS

Since $H_n^{(v)}(x)$ is Appell, we have

$$tH_n^{(v)}(x) = nH_{n-1}^{(v)}(x).$$

The recurrence formula is

$$H_{n+1}^{(v)}(x) = (x - vt)H_n^{(v)}(x)$$
$$= xH_n^{(v)}(x) - vnH_{n-1}^{(v)}(x). \qquad (4.2.2)$$

Applying the operator t to both sides and recalling that $tx - xt$ is the identity [Eq (3.6.3)], we get

$$(n + 1)H_n^{(v)}(x) = tH_{n+1}^{(v)}(x)$$
$$= txH_n^{(v)}(x) - vt^2 H_n^{(v)}(x)$$
$$= xtH_n^{(v)}(x) + H_n^{(v)}(x) - vt^2 H_n^{(v)}(x),$$

which simplifies to

$$vt^2 H_n^{(v)}(x) - xtH_n^{(v)}(x) + nH_n^{(v)}(x) = 0.$$

This is the familiar second-order differential equation for the Hermite polynomials.

One easily sees, by taking $p(x) = x^n$, that

$$(x - vt)p(x) = -e^{x^2/2v}(vt)e^{-x^2/2v}p(x),$$

and so by iterating the first equation in (4.2.2) we get the classical Rodrigues formula

$$H_n^{(v)}(x) = (-1)^n e^{x^2/2v}(vt)^n e^{-x^2/2v}.$$

EXPANSIONS

The polynomial expansion theorem is

$$p(x) = \sum_{k \geq 0} \frac{\langle e^{vt^2/2} \mid t^k p(x) \rangle}{k!} H_k^{(v)}(x).$$

If we take $p(x) = x^n$, then

$$x^n = \sum_{k \geq 0} \binom{n}{k} \langle e^{vt^2/2} \mid x^{n-k} \rangle H_k^{(v)}(x).$$

For $n = 2m$,

$$\langle e^{vt^2/2} \mid x^{2m-k} \rangle = \sum_{j=0}^{\infty} \left(\frac{v}{2}\right)^j \frac{1}{j!} \langle t^{2j} \mid x^{2m-k} \rangle$$

$$= \begin{cases} \left(\dfrac{v}{2}\right)^{m-1} \dfrac{(2m-2l)!}{(m-l)!} & \text{for } k = 2l, \\[2ex] 0 & \text{for } k \text{ odd,} \end{cases}$$

and so

$$x^{2m} = \sum_{l=0}^{m} \binom{2m}{2l} \frac{(2m-2l)!}{(m-l)!} \left(\frac{v}{2}\right)^{m-l} H_{2l}^{(v)}(x).$$

For $n = 2m + 1$,

$$\langle e^{vt^2/2} \mid x^{2m+1-k} \rangle = \sum_{j=0}^{\infty} \left(\frac{v}{2}\right)^j \frac{1}{j!} \langle t^{2j} \mid x^{2m+1-k} \rangle$$

$$= \begin{cases} \left(\dfrac{v}{2}\right)^{m-l} \dfrac{(2m-2l)!}{(m-l)!} & \text{for } k = 2l+1, \\[2ex] 0 & \text{for } k \text{ even,} \end{cases}$$

and so

$$x^{2m+1} = \sum_{l=0}^{m} \binom{2m+1}{2l+1} \frac{(2m-2l)!}{(m-l)!} \left(\frac{v}{2}\right)^{m-l} H_{2l+1}^{(v)}(x).$$

One of the nicest properties of the Hermite polynomials is expressed in the next lemma.

Lemma 4.2.1   For any polynomial $p(x)$,

$$\langle e^{t^2/2} \mid H_n(x)p(x) \rangle = \langle t^n e^{t^2/2} \mid p(x) \rangle$$

for all $n \geq 0$.

Proof   We verify this for $p(x) = x^m$ by induction on m. If $m = 0$, then

$$\langle e^{t^2/2} \mid H_n(x) \rangle = \delta_{n,0} = \langle t^n e^{t^2/2} \mid x^0 \rangle$$

for all $n \geq 0$. Suppose that for all $l \leq m$

$$\langle e^{t^2/2} \mid H_n(x)x^l \rangle = \langle t^n e^{t^2/2} \mid x^l \rangle$$

for all $n \geq 0$. Then we show that this also holds for $l = m + 1$. Now

$$\langle e^{t^2/2} \mid H_n(x)x^{m+1} \rangle = \langle \partial_t e^{t^2/2} \mid H_n(x)x^m \rangle$$
$$= \langle t e^{t^2/2} \mid H_n(x)x^m \rangle$$
$$= \langle e^{t^2/2} \mid t H_n(x)x^m \rangle$$
$$= \langle e^{t^2/2} \mid n H_{n-1}(x)x^m + m H_n(x)x^{m-1} \rangle$$
$$= \langle n t^{n-1} e^{t^2/2} \mid x^m \rangle + \langle m t^n e^{t^2/2} \mid x^{m-1} \rangle$$
$$= \langle n t^{n-1} e^{t^2/2} + t^{n+1} e^{t^2/2} \mid x^m \rangle$$
$$= \langle \partial_t t^n e^{t^2/2} \mid x^m \rangle$$
$$= \langle t^n e^{t^2/2} \mid x^{m+1} \rangle$$

for all $n \geq 0$. This completes the proof.

This lemma is most effective when used with the polynomial expansion theorem. If $p(x)$ is any polynomial, then

$$p(x)H_m(x) = \sum_{k \geq 0} \frac{1}{k!} \langle e^{t^2/2} \mid t^k p(x) H_m(x) \rangle H_k(x).$$

From the Leibniz formula and the lemma we have

$$\langle e^{t^2/2} \mid t^k p(x) H_m(x) \rangle = \left\langle e^{t^2/2} \middle| \sum_{j=0}^{k} \binom{k}{j} t^j p(x) t^{k-j} H_m(x) \right\rangle$$
$$= \left\langle e^{t^2/2} \middle| \sum_{j=0}^{k} \binom{k}{j} (m)_{k-j} [t^j p(x)] H_{m-k+j}(x) \right\rangle$$
$$= \sum_{j=0}^{k} \binom{k}{j} (m)_{k-j} \langle e^{t^2/2} \mid [t^j p(x)] H_{m-k+j}(x) \rangle$$
$$= \sum_{j=0}^{k} \binom{k}{j} (m)_{k-j} \langle t^{m-k+j} e^{t^2/2} \mid t^j p(x) \rangle$$
$$= \sum_{j=0}^{k} \binom{k}{j} (m)_{k-j} \langle e^{t^2/2} \mid t^{m-k+2j} p(x) \rangle,$$

and so, for any polynomial $p(x)$,

$$p(x)H_m(x) = \sum_{k \geq 0} \left[\sum_{j=0}^{k} \binom{m}{k-j} \frac{1}{j!} \langle e^{t^2/2} \mid t^{m-k+2j} p(x) \rangle \right] H_k(x). \qquad (4.2.3)$$

In particular, if $s_n(x)$ is an Appell sequence and we take $p(x) = s_n(x)$, then

$$t^{m-k+2j} s_n(x) = (n)_{m-k+2j} s_{n-m+k-2j}(x)$$
$$= (n)_j (n-j)_{m-k+j} s_{n-m+k-2j}(x),$$

and so

$$s_n(x)H_m(x) = \sum_{k\geq 0}\left[\sum_{j=0}^{k}\binom{m}{k-j}\binom{n}{j}(n-j)_{m-k+j}\langle e^{t^2/2} \mid s_{n-m+k-2j}(x)\rangle\right]H_k(x).$$

In case $s_n(x) = H_n(x)$,

$$\langle e^{t^2/2} \mid H_{n-m+k-2j}(x)\rangle = \delta_{n-m+k-2j,0}$$
$$= \delta_{n-m+k,2j},$$

and so for $n \geq m$,

$$H_n(x)H_m(x) = \sum_{\substack{k=n-m \\ 2j=n-m+k}}^{n+m}\binom{n}{j}\binom{m}{k-j}(n-j)!\,H_k(x).$$

This can be put into a more familiar form by observing that $n - m + k$ is even if and only if $n + m - k$ is even, and so we may set $n + m - k = 2l$. Then $2j = n - m + k = 2n - 2l$, and so $j = n - l$. Also, $k - j = m - l$ and we have

$$H_n(x)H_m(x) = \sum_{l=0}^{m}\binom{n}{l}\binom{m}{l}l!\,H_{n+m-2l}(x).$$

This well-known formula goes back to 1918.

MISCELLANEOUS RESULTS

The umbral composition of Hermite polynomials is relatively simple,

$$H_n^{(\nu)}\big(\mathbf{H}^{(\mu)}(x)\big) = e^{-\mu t^2/2}H_n^{(\nu)}(x)$$
$$= e^{-\mu t^2/2}e^{-\nu t^2/2}x^n$$
$$= e^{-(\nu+\mu)t^2/2}x^n$$
$$= H_n^{(\nu+\mu)}(x).$$

In particular,

$$H_n\big(\mathbf{H}(x)\big) = H_n^{(2)}(x) = 2^{n/2}H_n(x/2^{1/2}).$$

For $\mu = -\nu$ we get

$$H_n^{(\nu)}\big(\mathbf{H}^{(-\nu)}(x)\big) = H_n^{(0)}(x) = x^n,$$

and so $H_n^{(\nu)}(x)$ and $H_n^{(-\nu)}(x)$ are inverses under umbral composition. This nice behavior with respect to umbral composition is certainly not unique to the Hermite polynomials, and we shall speak briefly about this when we discuss cross sequences in Chapter 5.

We conclude with the multiplication theorem,

$$H_n^{(v)}(\alpha x) = \alpha^n \frac{g(t)}{g(t/\alpha)} H_n^{(v)}(x)$$

$$= \alpha^n e^{(1-\alpha^{-2})vt^2/2} H_n^{(v)}(x)$$

$$= \alpha^n H_n^{(v/\alpha^2)}(x),$$

which, of course, follows directly from the explicit expression of $H_n^{(v)}(x)$.

2.2. The Bernoulli Polynomials

DEFINITIONS

The Bernoulli polynomials $B_n^{(a)}(x)$ of order a form the Appell sequence for

$$g(t) = \left(\frac{e^t - 1}{t} \right)^a \qquad (a \neq 0).$$

We recall from Chapter 2 that

$$\left\langle \frac{e^t - 1}{t} \,\middle|\, p(x) \right\rangle = \int_0^1 p(u)\, du$$

and

$$\frac{e^t - 1}{t} p(x) = \int_x^{x+1} p(u)\, du.$$

The polynomials $B_n^{(1)}(x) = B_n(x)$ are by far the most common. The interested reader may consult Milne-Thomson [1] for a discussion of the higher order Bernoulli polynomials from a classical point of view.

It follows from Theorem 2.5.5 that

$$B_n^{(a)}(x) = \left(\frac{t}{e^t - 1} \right)^a x^n,$$

and so

$$\left(\frac{t}{e^t - 1} \right)^b B_n^{(a)}(x) = B_n^{(a+b)}(x).$$

A simple expression for the Bernoulli polynomials in terms of the lower factorials can be obtained from the polynomial expansion theorem,

$$B_n(x) = \sum_{k=0}^n \frac{\langle (e^t - 1)^k \,|\, B_n(x) \rangle}{k!} (x)_k.$$

Since for $k \geq 1$

$$\langle (e^t - 1)^k \,|\, B_n(x) \rangle = \left\langle (e^t - 1)^k \,\middle|\, \frac{t}{e^t - 1} x^n \right\rangle$$
$$= \langle (e^t - 1)^{k-1} \,|\, nx^{n-1} \rangle$$
$$= n(k - 1)! \, S(n - 1, k - 1),$$

we have

$$B_n(x) = B_n(0) + \sum_{k=1}^{n} \frac{n}{k} S(n - 1, k - 1)(x)_k.$$

GENERATING FUNCTION AND SHEFFER IDENTITY

The generating function of the Bernoulli polynomials is

$$\sum_{k=0}^{\infty} \frac{B_k^{(a)}(x)}{k!} t^k = \left(\frac{t}{e^t - 1} \right)^a e^{xt}.$$

Setting $x = 0$ gives the expansion

$$\left(\frac{t}{e^t - 1} \right)^a = \sum_{k=0}^{\infty} \frac{B_k^{(a)}(0)}{k!} t^k,$$

where the numbers $B_n^{(a)}(0)$ are the well-known Bernoulli numbers.
  The Appell identity is

$$B_n^{(a)}(x + y) = \sum_{k=0}^{n} \binom{n}{k} B_k^{(a)}(y) x^{n-k}.$$

Applying $[t/(e^t - 1)]^b$, we have

$$B_n^{(a+b)}(x + y) = \sum_{k=0}^{n} \binom{n}{k} B_k^{(a)}(x) B_{n-k}^{(b)}(y),$$

and for $a + b = 0$

$$(x + y)^n = \sum_{k=0}^{n} \binom{n}{k} B_k^{(a)}(x) B_{n-k}^{(-a)}(y).$$

Setting $y = 0$ in the Appell identity, we get

$$B_n^{(a)}(x) = \sum_{k=0}^{n} \binom{n}{k} B_{n-k}^{(a)}(0) x^k.$$

RECURRENCE RELATIONS AND OPERATIONAL FORMULAS

Since $B_n^{(a)}(x)$ is Appell,

$$t B_n^{(a)}(x) = n B_{n-1}^{(a)}(x). \tag{4.2.4}$$

Next we observe that

$$(e^t - 1)B_n^{(a)}(x) = (e^t - 1)\left(\frac{t}{e^t - 1}\right)^a x^n$$

$$= \left(\frac{t}{e^t - 1}\right)^{a-1} tx^n \qquad (4.2.5)$$

$$= nB_{n-1}^{(a-1)}(x),$$

which may be written as

$$B_n^{(a)}(x + 1) = B_n^{(a)}(x) + nB_{n-1}^{(a-1)}(x). \qquad (4.2.6)$$

In particular,

$$(e^t - 1)B_n(x) = nx^{n-1}$$

or

$$B_n(x + 1) = B_n(x) + nx^{n-1}.$$

Also,

$$\int_0^y B_n^{(a)}(u)\,du = \left\langle \frac{e^{yt} - 1}{t} \,\middle|\, B_n^{(a)}(x) \right\rangle$$

$$= \left\langle \frac{e^{yt} - 1}{t} \,\middle|\, \frac{1}{n+1}(e^t - 1)B_{n+1}^{(a+1)}(x) \right\rangle$$

$$= \frac{1}{n+1}\left\langle e^{yt} - 1 \,\middle|\, \frac{e^t - 1}{t} B_{n+1}^{(a+1)}(x) \right\rangle$$

$$= \frac{1}{n+1}\left\langle e^{yt} - 1 \,\middle|\, B_{n+1}^{(a)}(x) \right\rangle$$

$$= \frac{1}{n+1}\left[B_{n+1}^{(a)}(y) - B_{n+1}^{(a)}(0) \right].$$

The recurrence formula is

$$B_{n+1}^{(a)}(x) = \left(x - \frac{g'(t)}{g(t)} \right) B_n^{(a)}(x)$$

$$= \left[x - a\left(1 + \frac{1 + t - e^t}{t(e^t - 1)} \right) \right] B_n^{(a)}(x).$$

Applying t to both sides and recalling that $tx - xt$ is the identity, we get

$$(n + 1)B_n^{(a)}(x) = \left[xt + 1 - at - a\frac{1 + t - e^t}{e^t - 1} \right] B_n^{(a)}(x)$$

$$= (x - a)tB_n^{(a)}(x) + B_n^{(a)}(x) + aB_n^{(a)}(x) - a\frac{t}{e^t - 1}B_n^{(a)}(x)$$

$$= n(x - a)B_{n-1}^{(a)}(x) + B_n^{(a)}(x) + aB_n^{(a)}(x) - aB_n^{(a+1)}(x),$$

and so

$$B_n^{(a+1)}(x) = \left(1 - \frac{n}{a}\right)B_n^{(a)}(x) + n\left(\frac{x}{a} - 1\right)B_{n-1}^{(a)}(x), \qquad (4.2.7)$$

which expresses the Bernoulli polynomials of order $a + 1$ in terms of those of order a (for $a \neq 0$).

By the transfer formula the associated sequence for the delta series

$$f(t) = \left(\frac{e^t - 1}{t}\right)^a t$$

is

$$p_n(x) = x\left(\frac{f(t)}{t}\right)^{-n} x^{n-1}$$

$$= x\left(\frac{t}{e^t - 1}\right)^{na} x^{n-1}$$

$$= xB_{n-1}^{(na)}(x).$$

But if $a = 1$, then $f(t) = e^t - 1$ is the forward difference, and so we have

$$(x)_n = xB_{n-1}^{(n)}(x),$$

which gives

$$B_{n-1}^{(n)}(x) = x^{-1}(x)_n = (x - 1)_{n-1}.$$

Applying t^{n-k-1}, for $k \leq n - 1$, we obtain the operational formula

$$B_k^{(n)}(x) = \frac{k!}{(n-1)!}t^{n-k-1}(x - 1)_{n-1}. \qquad (4.2.8)$$

EXPANSIONS

The expansion theorem is

$$h(t) = \sum_{k=0}^{\infty} \frac{\langle h(t) | B_k^{(a)}(x)\rangle}{k!}\left(\frac{e^t - 1}{t}\right)^a t^k.$$

Taking $h(t) = e^{yt}$, we obtain the famous Euler–Maclaurin expansion,

$$e^{yt} = \sum_{k=0}^{\infty} \frac{B_k^{(a)}(y)}{k!} \left(\frac{e^t - 1}{t}\right)^a t^k.$$

Applying this to a polynomial $p(x)$ gives, for $a = 1$,

$$p(x + y) = \sum_{k \geq 0} \frac{B_k(y)}{k!} \int_x^{x+1} p^{(k)}(u)\, du.$$

Recalling the expansion of t^m in powers of the forward difference, we have

$$t^m = \sum_{k=m}^{\infty} \frac{\langle t^m \mid (x/a)_k \rangle}{k!} (e^{at} - 1)^k = \sum_{k=m}^{\infty} \frac{m!}{k!\, a^m} s(k, m) \left(\frac{e^{at} - 1}{t}\right)^k t^k.$$

Applying this to $B_n^{(a)}(x)$ gives

$$(n)_m B_{n-m}^{(a)}(x) = \sum_{k=m}^{n} \frac{m!}{k!\, a^m} s(k, m)(n)_k B_{n-k}^{(a-k)}(x) = \sum_{k=m}^{n} \binom{n}{k} \frac{m!}{a^m} s(k, m) B_{n-k}^{(a-k)}(x).$$

Since for any $m \geq 0$

$$\frac{e^t - 1}{t} = \sum_{k=m}^{\infty} \frac{1}{(k - m + 1)!} t^{k-m},$$

we have

$$t^m = \sum_{k=m}^{\infty} \frac{1}{(k - m + 1)!} \left(\frac{t}{e^t - 1}\right) t^k.$$

Applying this to x^n gives

$$(n)_m x^{n-m} = \sum_{k=m}^{n} \frac{(n)_k}{(k - m + 1)!} B_{n-k}(x).$$

For $m = 1$ this is the well-known formula

$$n x^{n-1} = \sum_{k=1}^{n} \binom{n}{k} B_{n-k}(x).$$

MISCELLANEOUS RESULTS

The multiplication theorem is

$$B_n(\alpha x) = \alpha^n \frac{g(t)}{g(t/\alpha)} B_n(x)$$

$$= \alpha^{n-1} \frac{e^t - 1}{e^{t/\alpha} - 1} B_n(x).$$

Now if $\alpha = m$ is a positive integer, then

$$B_n(mx) = m^{n-1} \frac{e^t - 1}{e^{t/m} - 1} B_n(x)$$

$$= m^{n-1} \sum_{k=0}^{m-1} e^{kt/m} B_n(x)$$

$$= m^{n-1} \sum_{k=0}^{m-1} B_n\left(x + \frac{k}{m}\right),$$

which is the classical multiplication theorem for Bernoulli polynomials.

From Proposition 2.1.11 we see that

$$\left\langle \left(\frac{e^{-t} - 1}{-t}\right)^a (-t)^k \,\middle|\, B_n^{(a)}(-x) \right\rangle = \left\langle \left(\frac{e^t - 1}{t}\right)^a t^k \,\middle|\, B_n^{(a)}(x) \right\rangle = n!\, \delta_{n,k},$$

and so the sequence

$$s_n(x) = B_n^{(a)}(-x)$$

is Sheffer for

$$\left(\left(\frac{1 - e^{-t}}{t}\right)^a, -t\right).$$

Therefore, since $p_n(x) = (-x)^n$ is associated to $f(t) = -t$,

$$B_n^{(a)}(-x) = \left(\frac{t}{1 - e^{-t}}\right)^a (-x)^n$$

$$= e^{at}\left(\frac{t}{e^t - 1}\right)^a (-x)^n$$

$$= (-1)^n e^{at} B_n^{(a)}(x)$$

$$= (-1)^n B_n^{(a)}(x + a).$$

This is known as the complementary argument theorem.

Let us discuss briefly the Bernoulli numbers. Taking $x = 0$ in (4.2.6) gives

$$B_n^{(a)}(1) = B_n^{(a)}(0) + n B_{n-1}^{(a-1)}(0).$$

Taking $x = 0$ in (4.2.7), with a replaced by $a - 1$, gives

$$B_n^{(a)}(0) = \left(1 - \frac{n}{a - 1}\right) B_n^{(a-1)}(0) - n B_{n-1}^{(a-1)}(0).$$

Combining these two formulas, we obtain

$$B_n^{(a)}(1) = \left(1 - \frac{n}{a - 1}\right) B_n^{(a-1)}(0).$$

The operational formula (4.2.8) is equivalent to

$$t^m(x)_n = \frac{n!}{(n-m)!} B_{n-m}^{(n+1)}(x+1)$$

for $m \geq 1$. This leads to a connection between the Bernoulli numbers and the Stirling numbers of the first kind,

$$s(n,m) = \frac{1}{m!} \langle t^m \mid (x)_n \rangle$$

$$= \frac{1}{m!} \langle t^0 \mid t^m(x)_n \rangle$$

$$= \binom{n}{m} B_{n-m}^{(n+1)}(1)$$

$$= \binom{n-1}{m-1} B_{n-m}^{(n)}(0).$$

Thus

$$(x)_n = \sum_{k=0}^{n} \binom{n-1}{m-1} B_{n-k}^{(n)}(0) x^k.$$

We may connect the Bernoulli numbers to the Stirling numbers of the second kind as follows:

$$S(n,k) = \frac{1}{k!} \langle (e^t - 1)^k \mid x^n \rangle$$

$$= \frac{1}{k!} \left\langle \left(\frac{e^t - 1}{t} \right)^k \middle| t^k x^n \right\rangle$$

$$= \binom{n}{k} \left\langle t^0 \middle| \left(\frac{t}{e^t - 1} \right)^{-k} x^{n-k} \right\rangle$$

$$= \binom{n}{k} B_{n-k}^{(-k)}(0),$$

and so

$$x^n = \sum_{k=0}^{n} \binom{n}{k} B_{n-k}^{(-k)}(0)(x)_k.$$

One of the most appealing aspects of the Bernoulli numbers is their connection with the partial sums of the harmonic series. If we let

$$h_n = 1 + \frac{1}{2} + \frac{1}{3} + \cdots + \frac{1}{n},$$

then since

$$x^{(n+1)} = x(x+1)(x+2)\cdots(x+n)$$

$$= n!\, x(x+1)\left(\frac{x}{2}+1\right)\cdots\left(\frac{x}{n}+1\right),$$

we have

$$h_n = \left\langle \frac{t^2}{2} \,\middle|\, \frac{1}{n!}x^{(n+1)} \right\rangle.$$

But $p_n(x) = x^{(n)}$ is associated to the backward difference functional

$$f(t) = 1 - e^{-t},$$

as can easily be seen by the recurrence formula. Thus the transfer formula gives

$$h_n = \left\langle \frac{t^2}{2} \,\middle|\, \frac{1}{n!}x\left(\frac{1-e^{-t}}{t}\right)^{-n-1} x^n \right\rangle$$

$$= \frac{1}{n!}\left\langle t \,\middle|\, \left(\frac{t}{1-e^{-t}}\right)^{n+1} x^n \right\rangle$$

$$= \frac{1}{n!}\left\langle t \,\middle|\, \left(\frac{1}{e^{-t}}\right)^{n+1}\left(\frac{t}{e^t-1}\right)^{n+1} x^n \right\rangle$$

$$= \frac{1}{n!}\langle t \,|\, e^{(n+1)t} B_n^{(n+1)}(x)\rangle$$

$$= \frac{1}{n!}\langle e^{(n+1)t} \,|\, t B_n^{(n+1)}(x)\rangle$$

$$= \frac{1}{(n-1)!} B_{n-1}^{(n+1)}(n+1).$$

Using the complementary argument theorem, we finally get

$$1 + \frac{1}{2} + \cdots + \frac{1}{n} = \frac{(-1)^{n-1}}{(n-1)!} B_{n-1}^{(n+1)}(0).$$

2.3. The Euler Polynomials

DEFINITIONS

The Euler polynomials $E_n^{(a)}(x)$ of order a form the Appell sequence for

$$g(t) = \left(\frac{e^t+1}{2}\right)^a \qquad (a \neq 0).$$

The properties of the Euler polynomials are similar in spirit to those of the Bernoulli polynomials. We write $E_n^{(1)}(x) = E_n(x)$.

It follows from Theorem 2.5.5 that

$$E_n^{(a)}(x) = \left(\frac{2}{e^t + 1} \right)^a x^n,$$

and so

$$\left(\frac{2}{e^t + 1} \right)^b E_n^{(a)}(x) = E_n^{(a+b)}(x).$$

Since

$$\left(\frac{2}{e^t + 1} \right)^a = \left(1 \Big/ \left(1 + \frac{e^t - 1}{2} \right) \right)^a = \sum_{j=0}^{\infty} \binom{-a}{j} \left(\frac{e^t - 1}{2} \right)^j,$$

we have the expression

$$E_n^{(a)}(x) = \sum_{j=0}^{n} \binom{-a}{j} \frac{1}{2^j} (e^t - 1)^j x^n. \tag{4.2.9}$$

From Theorem 2.3.8 we get

$$(e^t - 1)^j x^n = \sum_{k=0}^{n} \frac{1}{k!} \langle (e^t - 1)^j \mid x^k \rangle x^{n-k}$$

$$= \sum_{k=0}^{n} \frac{j!}{k!} S(k, j) x^{n-k},$$

and so

$$E_n^{(a)}(x) = \sum_{k=0}^{n} \left[\sum_{j=0}^{n} \binom{-a}{j} \frac{j!}{k!} \frac{1}{2^j} S(k, j) \right] x^{n-k}. \tag{4.2.10}$$

Equation (4.2.9) gives another formula for $E_n^{(a)}(x)$. Since

$$x^n = \sum_{k=0}^{n} S(n, k)(x)_k,$$

we have

$$(e^t - 1)^j x^n = \sum_{k=j}^{n} S(n, k)(k)_j (x)_{k-j},$$

and so

$$E_n^{(a)}(x) = \sum_{j=0}^{n} \sum_{k=j}^{n} \binom{-a}{j} \frac{1}{2^j} (k)_j S(n, k)(x)_{k-j}.$$

This is known as the Newton expansion of the Euler polynomials.

Since $p_n(x) = (x - a/2)^n$ is Appell for the series $e^{at/2}$, the expansion of $E_n^{(a)}(x)$ in terms of $p_n(x)$ is

$$E_n^{(a)}(x) = \sum_{k=0}^{n} \frac{\langle e^{at/2} t^k \mid E_n^{(a)}(x) \rangle}{k!} \left(x - \frac{a}{2} \right)^k$$

$$= \sum_{k=0}^{n} \binom{n}{k} \langle e^{at/2} \mid E_{n-k}^{(a)}(x) \rangle \left(x - \frac{a}{2} \right)^k$$

$$= \sum_{k=0}^{n} \binom{n}{k} E_k^{(a)} \left(\frac{a}{2} \right) \left(x - \frac{a}{2} \right)^{n-k}.$$

The numbers $2^n E_n^{(a)}(a/2)$ are known as the Euler numbers, in contradistinction to the Bernoulli case. (This terminology is due to Nörlund.)

GENERATING FUNCTION AND SHEFFER IDENTITY

The generating function for the Euler polynomials is

$$\sum_{k=0}^{\infty} \frac{E_k^{(a)}(x)}{k!} t^k = \left(\frac{2}{e^t + 1} \right)^a e^{xt}.$$

Incidentally, if we set $a = 1$, $t = 2iu$, and $x = 0$, we get

$$\sum_{k=0}^{\infty} \frac{2^k E_k(0)}{k!} (iu)^k = \frac{2}{e^{2iu} + 1} = 1 - i \tan u,$$

and if we set $a = 1$, $t = 2iu$, and $x = \frac{1}{2}$, then

$$\sum_{k=0}^{\infty} \frac{2^k E_k(\frac{1}{2})}{k!} (iu)^k = \frac{2}{e^{2iu} + 1} e^{iu} = \frac{2}{e^{iu} + e^{-iu}} = \sec u.$$

For this reason the numbers $2^k E_k^{(a)}(0)$ and $2^k E_k(\frac{1}{2})$ are known as the tangent and secant coefficients, respectively.

The Appell identity is

$$E_n^{(a)}(x + y) = \sum_{k=0}^{n} \binom{n}{k} E_k^{(a)}(y) x^{n-k},$$

and as before we have

$$E_n^{(a+b)}(x + y) = \sum_{k=0}^{n} \binom{n}{k} E_k^{(a)}(y) E_{n-k}^{(b)}(x)$$

and

$$(x + y)^n = \sum_{k=0}^{n} \binom{n}{k} E_k^{(a)}(y) E_{n-k}^{(-a)}(x).$$

Setting $y = 0$ in the Appell identity gives

$$E_n^{(a)}(x) = \sum_{k=0}^{n} \binom{n}{k} E_{n-k}^{(a)}(0) x^k.$$

RECURRENCE RELATIONS AND OPERATIONAL FORMULAS

Since $E_n^{(a)}(x)$ is Appell,

$$t E_n^{(a)}(x) = n E_{n-1}^{(a)}(x).$$

Further, we have

$$(e^t + 1) E_n^{(a)}(x) = (e^t + 1) \left(\frac{2}{e^t + 1} \right)^a x^n$$

$$= 2 \left(\frac{2}{e^t + 1} \right)^{a-1} x^n = 2 E_n^{(a-1)}(x),$$

and so

$$E_n^{(a)}(x + 1) = 2 E_n^{(a-1)}(x) - E_n^{(a)}(x). \tag{4.2.11}$$

Next we have

$$\int_0^y E_n^{(a)}(u)\, du = \left\langle \frac{e^{yt} - 1}{t} \middle| E_n^{(a)}(x) \right\rangle$$

$$= \left\langle \frac{e^{yt} - 1}{t} \middle| \frac{1}{n+1} t E_{n+1}^{(a)}(x) \right\rangle$$

$$= \frac{1}{n+1} \langle e^{yt} - 1 \mid E_{n+1}^{(a)}(x) \rangle$$

$$= \frac{1}{n+1} [E_{n+1}^{(a)}(y) - E_{n+1}^{(a)}(0)].$$

The recurrence formula is

$$E_{n+1}^{(a)}(x) = \left(x - \frac{g'(t)}{g(t)} \right) E_n^{(a)}(x) = \left(x - \frac{ae^t}{e^t + 1} \right) E_n^{(a)}(x)$$

$$= x E_n^{(a)}(x) - \frac{a}{2} E_n^{(a+1)}(x + 1)$$

or

$$\frac{a}{2} E_n^{(a+1)}(x + 1) = x E_n^{(a)}(x) - E_n^{(a)}(x) - E_{n+1}^{(a)}(x).$$

Combining this with (4.2.11), we get the recurrence

$$E_n^{(a+1)}(x) = \frac{2}{a}E_{n+1}^{(a)} + 2\left(1 - \frac{x}{a}\right)E_n^{(a)}(x).$$

EXPANSIONS

The expansion theorem is

$$h(t) = \sum_{k=0}^{\infty} \frac{\langle h(t) \mid E_k^{(a)}(x)\rangle}{k!}\left(\frac{e^t + 1}{2}\right)^a t^k.$$

Taking $h(t) = e^{yt}$, we get Boole's summation formula

$$e^{yt} = \sum_{k=0}^{\infty} \frac{E_k^{(a)}(y)}{k!}\left(\frac{e^t + 1}{2}\right)^a t^k.$$

Applying this to $p(x)$ for $a = 1$ gives

$$p(x + y) = \sum_{k \geq 0} \frac{E_k^{(a)}(x)}{k!}\frac{p^{(k)}(1) + p^{(k)}(0)}{2}.$$

Since for $m \geq 0$

$$e^t + 1 = \sum_{k=m}^{\infty}\left(\delta_{k,m} + \frac{1}{(k-m)!}\right)t^{k-m},$$

we have

$$t^m = \sum_{k=m}^{\infty}\left(\delta_{k,m} + \frac{1}{(k-m)!}\right)\frac{1}{2}\left(\frac{2}{e^t + 1}\right)t^k.$$

Applying this to x^n gives

$$(n)_m x^{n-m} = \sum_{k=m}^{\infty}\left(\delta_{k,m} + \frac{1}{(k-m)!}\right)\frac{(n)_k}{2}E_{n-k}(x),$$

and for $m = 1$

$$nx^{n-1} = \sum_{k=1}^{\infty}\left(\delta_{k,1} + \frac{1}{(k-1)!}\right)\frac{(n)_k}{2}E_{n-k}(x).$$

Theorem 2.5.9 gives us

$$xE_n^{(a)}(x) = E_{n+1}^{(a)}(x) + \sum_{k=0}^{n}\binom{n}{k}\left\langle \frac{a}{2}e^t\left(\frac{e^t + 1}{2}\right)^{a-1}\middle| E_{n-k}^{(a)}(x)\right\rangle E_k^{(a)}(x)$$

$$= E_{n+1}^{(a)}(x) + \sum_{k=0}^{n}\binom{n}{k}\frac{a}{2}E_{n-k}(1)E_k^{(a)}(x).$$

MISCELLANEOUS RESULTS

The multiplication theorem is

$$E_n(\alpha x) = \alpha^n \frac{1 + e^t}{1 + e^{t/\alpha}} E_n(x).$$

If $\alpha = 2m + 1$ is a positive odd integer, then

$$\sum_{k=0}^{2m} [-e^{t/(2m+1)}]^k = \frac{1 - [-e^{t/(2m+1)}]^{2m+1}}{1 - [-e^{t/(2m+1)}]}$$

$$= \frac{1 + e^t}{1 + e^{t/(2m+1)}},$$

and so

$$E_n((2m + 1)x) = \alpha^n \sum_{k=0}^{2m} (-1)^k e^{kt/(2m+1)} E_n(x)$$

$$= \alpha^n \sum_{k=0}^{2m} (-1)^k E_n\left(x + \frac{k}{2m + 1}\right).$$

For $\alpha = 2m$ a positive integer we must be a bit more enterprising and write

$$E_n(2mx) = (2m)^n \frac{1 + e^t}{1 + e^{t/2m}} E_n(x)$$

$$= 2(2m)^n \frac{1}{1 + e^{t/2m}} x^n$$

$$= -2(2m)^n \left(\frac{1 - e^t}{1 + e^{t/2m}}\right)\left(\frac{t}{e^t - 1}\right)\frac{1}{n + 1} x^{n+1}$$

$$= -\frac{2(2m)^n}{n + 1}\left(\frac{1 - e^t}{1 + e^{t/2m}}\right) B_{n+1}(x)$$

$$= -\frac{2(2m)^n}{n + 1} \sum_{k=0}^{2m-1} [-e^{t/2m}]^k B_{n+1}(x)$$

$$= -\frac{2(2m)^n}{n + 1} \sum_{k=0}^{2m-1} (-1)^k B_{n+1}\left(x + \frac{k}{2m}\right).$$

Setting $m = 1$ gives a nice formula connecting the Euler polynomials with the Bernoulli polynomials,

$$E_n(2x) = -\frac{2^{n+1}}{n + 1}\left[B_{n+1}(x) - B_{n+1}\left(x + \frac{1}{2}\right)\right].$$

By the multiplication theorem for Bernoulli polynomials

$$B_{n+1}\left(x + \frac{1}{2}\right) = \frac{1}{2^n}B_{n+1}(2x) - B_{n+1}(x),$$

and so

$$E_n(2x) = \frac{2}{n+1}[B_{n+1}(2x) - 2^{n+1}B_{n+1}(x)].$$

Since $E_n^{(a)}(-x)$ is Sheffer for

$$\left(\left(\frac{e^{-t}+1}{2}\right)^a, -t\right),$$

we have

$$E_n^{(a)}(-x) = \left(\frac{2}{e^{-t}+1}\right)^a(-x)^n$$

$$= (-1)^n e^{at}\left(\frac{2}{1+e^t}\right)^a x^n$$

$$= (-1)^n E_n^{(a)}(x+a),$$

which is the complementary argument theorem for the Euler polynomials. By the transfer formula the associated sequence for the delta series

$$f(t) = \frac{e^{-at/2}(e^t+1)^a}{2^a}t$$

$$= t\left[\cosh\frac{t}{2}\right]^a$$

is

$$p_n(x) = x\left(\frac{f(t)}{t}\right)^{-n}x^{n-1}$$

$$= xe^{ant/2}\left(\frac{2}{e^t+1}\right)^{an}x^{n-1}$$

$$= xe^{ant/2}E_{n-1}^{(an)}(x)$$

$$= xE_{n-1}^{(an)}\left(x + \frac{an}{2}\right).$$

This fact can be used to derive additional properties of the Euler polynomials.

3. SHEFFER SEQUENCES

For ease of reference we list the key results. Let $s_n(x)$ be Sheffer for $(g(t), f(t))$.

Generating function

$$\sum_{k=0}^{\infty} \frac{s_k(x)}{k!} t^k = \frac{1}{g(\bar{f}(t))} e^{x\bar{f}(t)}.$$

Sheffer identity

$$s_n(x + y) = \sum_{k=0}^{n} \binom{n}{k} s_k(x) p_{n-k}(y).$$

Theorem 2.3.6

$$s_n(x) = g(t)^{-1} p_n(x).$$

Theorem 2.3.7

$$f(t) s_n(x) = n s_{n-1}(x).$$

Recurrence formula

$$s_{n+1}(x) = \left(x - \frac{g'(t)}{g(t)} \right) \frac{1}{f'(t)} s_n(x).$$

Expansion theorem

$$h(t) = \sum_{k=0}^{\infty} \frac{\langle h(t) \mid s_k(x) \rangle}{k!} g(t) f(t)^k.$$

Polynomial expansion theorem

$$p(x) = \sum_{k \geq 0} \frac{\langle g(t) f(t)^k \mid p(x) \rangle}{k!} s_k(x).$$

Theorem 3.5.6

$$\langle h(t) \mid q(s(x)) \rangle = \langle g(\bar{f}(t))^{-1} h(\bar{f}(t)) \mid q(x) \rangle,$$
$$\langle h(t) \mid s_n(x) \rangle = \langle g(\bar{f}(t))^{-1} h(\bar{f}(t)) \mid x^n \rangle.$$

Theorem 2.3.11

$$x s_n(x) = \sum_{k=0}^{n+1} \left[\binom{n}{k} \langle g'(t) \mid s_{n-k}(x) \rangle \right.$$
$$\left. + \binom{n}{k-1} \langle g(t) f'(t) \mid s_{n-k+1}(x) \rangle \right] s_k(x).$$

Theorem 2.3.12

$$s'_n(x) = \sum_{k=0}^{n-1}\binom{n}{k}\langle t \mid p_{n-k}(x)\rangle s_k(x)$$

or

$$s'_n(x) = \sum_{k=0}^{n-1}\binom{n}{k}\langle \bar{f}(t) \mid x^{n-k}\rangle s_k(x).$$

Conjugate representation

$$s_n(x) = \sum_{k=0}^{n}\frac{1}{k!}\langle g(\bar{f}(t))^{-1}\bar{f}(t)^k \mid x^n\rangle x^k.$$

Proposition 2.1.11

$$\langle h(t) \mid p(ax)\rangle = \langle h(at) \mid p(x)\rangle.$$

Theorem 2.1.10

$$\langle h(t) \mid xp(x)\rangle = \langle \partial_t h(t) \mid p(x)\rangle.$$

3.1. The Laguerre Polynomials

DEFINITIONS

The Laguerre polynomials $L_n^{(\alpha)}(x)$ of order α form the Sheffer sequence for the pair

$$g(t) = (1 - t)^{-\alpha-1}, \qquad f(t) = t/(t-1).$$

Evidentally, the sequence $L_n^{(-1)}(x) = L_n(x)$ is associated to the delta series $f(t)$. However, the polynomials $L_n^{(0)}(x)$ are generally known as the (simple) Laguerre polynomials.

Once again there is some variation in the definition of Laguerre polynomials. Many sources, including Erdelyi [1], take the Laguerre polynomials to be $L_n^{(\alpha)}(x)/n!$.

We note that

$$f'(t) = -1/(1 - t)^2 \qquad \text{and} \qquad \bar{f}(t) = t/(t-1) = f(t).$$

From Theorem 2.3.6 we see that

$$L_n^{(\alpha)}(x) = (1 - t)^{\alpha+1}L_n(x),$$

and so

$$(1 - t)^\beta L_n^{(\alpha)}(x) = L_n^{(\alpha+\beta)}(x). \tag{4.3.1}$$

The transfer formula gives an explicit expression for the sequence $L_n(x)$,

$$L_n(x) = f'(t)\left(\frac{f(t)}{t}\right)^{-n-1} x^n$$

$$= -(1-t)^2(t-1)^{n+1}x^n$$

$$= (-1)^n(1-t)^{n-1}x^n$$

$$= (-1)^n \sum_{k=0}^{n-1}\binom{n-1}{k}(-1)^k t^k x^n$$

$$= \sum_{k=0}^{n-1}\binom{n-1}{k}(-1)^{n-k}(n)_k x^{n-k}$$

$$= \sum_{k=1}^{n}\binom{n-1}{k-1}\frac{n!}{k!}(-x)^k.$$

Incidentally, the coefficients of $(-x)^k$ are the Lah numbers $B_{n,k}(1!, 2!, 3!, \ldots)$ that we encountered in Section 1.8.

From this and (4.3.1) we get

$$L_n^{(\alpha)}(x) = (1-t)^{\alpha+1}L_n(x)$$

$$= (-1)^n(1+t)^{\alpha+n}x^n$$

$$= \sum_{k=0}^{n}\binom{n+\alpha}{n-k}\frac{n!}{k!}(-x)^k.$$

It is worth noting separately that

$$L_n^{(\alpha)}(x) = (-1)^n(1-t)^{\alpha+n}x^n.$$

To derive the classical Rodrigues formula we observe that, as operators,

$$1 - t = -e^x t e^{-x},$$

and so

$$(1-t)^n = (-1)^n e^x t^n e^{-x}.$$

Furthermore, it is easy to see that

$$x^{-\alpha}(1-t)^n x^{\alpha+n} = (1-t)^{\alpha+n}x^n,$$

and so

$$L_n^{(\alpha)}(x) = (-1)^n(1-t)^{\alpha+n}x^n$$

$$= (-1)^n x^{-\alpha}(1-t)^n x^{\alpha+n}$$

$$= x^{-\alpha}e^x t^n x^{-x}e^{\alpha+n}.$$

GENERATING FUNCTION AND SHEFFER IDENTITY

The generating function for the Laguerre polynomials is

$$\sum_{k=0}^{\infty} \frac{L_k^{(\alpha)}(x)}{k!} t^k = (1 - t)^{-\alpha - 1} e^{xt/(t-1)}.$$

The Sheffer identity is

$$L_n^{(\alpha)}(x + y) = \sum_{k=0}^{n} \binom{n}{k} L_k^{(\alpha)}(y) L_{n-k}(x).$$

Applying $(1 - t)^{\beta + 1}$ gives

$$L_n^{(\alpha + \beta + 1)}(x + y) = \sum_{k=0}^{n} \binom{n}{k} L_k^{(\alpha)}(y) L_{n-k}^{(\beta)}(x).$$

RECURRENCE RELATIONS AND OPERATIONAL FORMULAS

Theorem 2.3.7 is

$$\frac{t}{t - 1} L_n^{(\alpha)}(x) = n L_{n-1}^{(\alpha)}(x), \tag{4.3.2}$$

which is equivalent to

$$t L_n^{(\alpha)}(x) = nt L_{n-1}^{(\alpha)}(x) - n L_{n-1}^{(\alpha)}(x)$$

and to

$$t L_n^{(\alpha)}(x) = -n L_{n-1}^{(\alpha+1)}(x).$$

The recurrence formula is

$$
\begin{aligned}
L_{n+1}^{(\alpha)}(x) &= \left(x - \frac{g'(t)}{g(t)} \right) \frac{1}{f'(t)} L_n^{(\alpha)}(x) \\
&= -[x - (\alpha + 1)(1 - t)^{-1}](1 - t)^2 L_n^{(\alpha)}(x) \\
&= (xt - x + \alpha + 1)(1 - t) L_n^{(\alpha)}(x) \\
&= (xt - x + \alpha + 1) L_n^{(\alpha+1)}(x). \tag{4.3.3}
\end{aligned}
$$

From this we see that

$$L_{n+m}^{(\alpha)}(x) = (xt - x + \alpha + 1)^{(n)} L_m^{(\alpha+m)}(x),$$

and if $m = 0$, we have a formula of Carlitz,

$$L_n^{(\alpha)}(x) = (xt - x + \alpha + 1)^{(n)} 1.$$

From (4.3.2) and the recurrence formula we easily obtain the familiar second-order differential equation for the Laguerre polynomials. Writing

(4.3.2) as

$$(1 - t)L_{n-1}^{(\alpha)}(x) = -(1/n)tL_n^{(\alpha)}(x)$$

and substituting into (4.3.3), where $n + 1$ has been replaced by n, we get

$$L_n^{(\alpha)}(x) = -(1/n)(xt - x + \alpha + 1)tL_n^{(\alpha)}(x),$$

which simplifies to

$$[xt^2 + (\alpha + 1 - x)t + n]L_n^{(\alpha)}(x) = 0.$$

EXPANSIONS

Theorem 2.3.11 is

$$xL_n^{(\alpha)}(x) = \sum_{k=0}^{n+1}\left[\binom{n}{k}\langle g'(t) \mid L_{n-k}^{(\alpha)}(x)\rangle \right.$$
$$\left. + \binom{n}{k-1}\langle g(t)f'(t) \mid L_{n-k+1}^{(\alpha)}(x)\rangle \right]L_k^{(\alpha)}(x).$$

But

$$\langle g'(t) \mid L_m^{(\alpha)}(x)\rangle = (\alpha + 1)\langle(1 - t)^{-\alpha-2} \mid L_m^{(\alpha)}(x)\rangle = (\alpha + 1)\langle t \mid L_m^{(-2)}(x)\rangle$$
$$= (\alpha + 1)\binom{m-2}{m}m! = (\alpha + 1)(\delta_{m,0} - \delta_{m,1})$$

and

$$\langle g(t)f'(t) \mid L_m^{(\alpha)}(x)\rangle = \langle-(1 - t)^{-\alpha-3} \mid L_m^{(\alpha)}(x)\rangle = -\langle t \mid L_m^{(-3)}(x)\rangle$$
$$= -\binom{m-3}{m}m! = -(\delta_{m,0} - 2\delta_{m,1} + 2\delta_{m,2}).$$

Thus

$$xL_n^{(\alpha)}(x) = \sum_{k=0}^{n+1}\left[\binom{n}{k}(\alpha + 1)(\delta_{n,k} - \delta_{n-1,k}) \right.$$
$$\left. + \binom{n}{k-1}(-\delta_{n+1,k} + 2\delta_{n,k} - 2\delta_{n-1,k})\right]L_k^{(\alpha)}(x)$$
$$= -L_{n+1}^{(\alpha)}(x) + (\alpha + 1 + 2n)L_n^{(\alpha)}(x)$$
$$-(n(\alpha + 1) + n(n - 1))L_{n-1}^{(\alpha)}(x),$$

which is the recurrence

$$L_{n+1}^{(\alpha)}(x) - (2n + \alpha + 1 - x)L_n^{(\alpha)}(x) + n(n + \alpha)L_{n-1}^{(\alpha)}(x) = 0.$$

Theorem 2.3.12 gives

$$tL_n^{(\alpha)}(x) = \sum_{k=0}^{n-1}\binom{n}{k}\langle t \mid L_{n-k}(x)\rangle L_k^{(\alpha)}(x)$$

$$= -\sum_{k=0}^{n-1}\binom{n}{k}(n-k)!\,L_k^{(\alpha)}(x) = -\sum_{k=0}^{n-1}\frac{n!}{k!}L_k^{(\alpha)}(x).$$

The polynomial expansion theorem is

$$p(x) = \sum_{k\geq 0}\frac{1}{k!}\left\langle (1-t)^{-\alpha-1}\left(\frac{t}{t-1}\right)^k \middle| p(x)\right\rangle L_k^{(\alpha)}(x).$$

Let us set $p(x) = xtL_n^{(\alpha)}(x)$. Then, since

$$\left\langle (1-t)^{-\alpha-1}\left(\frac{t}{t-1}\right)^k \middle| xtL_n^{(\alpha)}(x)\right\rangle$$

$$= \langle(-1)^k(1-t)^{-\alpha-1-k}t^k \mid xtL_n^{(\alpha)}(x)\rangle$$

$$= \langle(-1)^k(1-t)^{-\alpha-1-k}kt^{k-1}$$

$$\quad + (-1)^k(\alpha+1+k)(1-t)^{-\alpha-2-k}t^k \mid tL_n^{(\alpha)}(x)\rangle$$

$$= \left\langle k(1-t)^{-\alpha-1}\left(\frac{t}{t-1}\right)^k\right.$$

$$\left. - (\alpha+1+k)(1-t)^{-\alpha-1}\left(\frac{t}{t-1}\right)^{k+1}\middle| L_n^{(\alpha)}(x)\right\rangle$$

$$= kn!\,\delta_{n,k} - (\alpha+1+k)n!\,\delta_{n,k+1}$$

$$= nn!\,\delta_{n,k} - (\alpha+n)n!\,\delta_{n-1,k},$$

we have

$$xtL_n^{(\alpha)}(x) = nL_n^{(\alpha)}(x) - n(\alpha+n)L_{n-1}^{(\alpha)}(x).$$

This is just a small sample of the many formulas involving the Laguerre polynomials which can easily be obtained by umbral methods.

MISCELLANEOUS RESULTS

Let us discuss briefly the umbral composition of Laguerre polynomials. To compute $L_n^{(\alpha)}(\mathbf{L}^{(\beta)}(x))$ we employ Theorem 3.5.5, where

$$(g(t), f(t)) = ((1-t)^{-\beta-1}, t/(t-1))$$

and

$$(h(t), l(t)) = ((1-t)^{-\alpha-1}, t/(t-1)).$$

Then, according to this theorem, $L_n^{(\alpha)}(\mathbf{L}^{(\beta)}(x))$ is Sheffer for

$$\left(g(t)h(f(t)), l(f(t))\right) = \left((1 - t)^{\alpha - \beta}, t\right),$$

and so

$$L_n^{(\alpha)}(\mathbf{L}^{(\beta)}(x)) = (1 - t)^{\beta - \alpha} x^n = (-1)^n L_n^{(\beta - \alpha - n)}(x).$$

If we set $\alpha = \beta$, then

$$L_n^{(\alpha)}(\mathbf{L}^{(\alpha)}(x)) = x^n,$$

which shows that the Laguerre polynomials are self-inverse under umbral composition.

If we let

$$s_n^{(\lambda)}(x) = (1 - t)^{-\lambda} x^n,$$

then

$$s_n^{(\lambda)}(x) = (-1)^n L_n^{(-\lambda - n)}(x)$$

and

$$L_n^{(\alpha)}\left(L^{(\beta)}(x)\right) = s_n^{(\alpha - \beta)}(x). \tag{4.3.4}$$

Furthermore, since $s_n^{(\lambda)}(x)$ is Sheffer for $((1 - t)^\lambda, t)$, another application of Theorem 3.5.5 shows that $s_n^{(\lambda)}(\mathbf{L}^{(\mu)}(x))$ is Sheffer for $((1 - t)^{-\lambda - \mu - 1}, t/(t - 1))$. Thus

$$s_n^{(\lambda)}\left(\mathbf{L}^{(\mu)}(x)\right) = L_n^{(\lambda + \mu)}(x). \tag{4.3.5}$$

Combining (4.3.4) and (4.3.5), we obtain Rota's umbral composition law for Laguerre polynomials as

$$L_n^{(\alpha_1)}\left(\mathbf{L}^{(\alpha_2)}(\mathbf{L}^{(\alpha_3)}(\cdots(\mathbf{L}^{(\alpha_k)}(x))\cdots)))\right)$$

$$= \begin{cases} L_n^{(\alpha_1 - \alpha_2 + \alpha_3 - \cdots + \alpha_k)}(x), & k \quad \text{odd}, \\ s_n^{(\alpha_1 - \alpha_2 + \alpha_3 - \cdots - \alpha_k)}(x), & k \quad \text{even}. \end{cases}$$

3.2. The Bernoulli Polynomials of the Second Kind

DEFINITIONS

The Bernoulli polynomials $b_n(x)$ of the second kind form the Sheffer sequence for

$$g(t) = \frac{t}{e^t - 1},$$

$$f(t) = e^t - 1.$$

We recall that

$$\frac{e^t - 1}{t} p(x) = \int_x^{x+1} p(u)\, du$$

and

$$\left\langle \frac{e^t - 1}{t} \middle| p(x) \right\rangle = \int_0^1 p(u)\, du.$$

Jordan [1] makes a study of these polynomials, but he defines them as $b_n(x)/n!$. Theorem 2.3.6 is

$$b_n(x) = \frac{e^t - 1}{t} (x)_n$$

$$= \int_x^{x+1} (u)_n\, du.$$

Also,

$$t b_n(x) = (e^t - 1)(x)_n = n(x)_{n-1},$$

and from this we get

$$b_n(x) - b_n(0) = n \int_0^x (u)_{n-1}\, du. \qquad (4.3.6)$$

Theorem 2.3.7 is

$$(e^t - 1) b_n(x) = n b_{n-1}(x)$$

or

$$b_n(x + 1) = b_n(x) + n b_{n-1}(x).$$

Setting $x = 0$ gives

$$b_n(1) = b_n(0) + n b_{n-1}(0).$$

If we combine this with (4.3.6), we get

$$b_n(0) = \int_0^1 (u)_n\, du.$$

Let us call the numbers

$$b_n(0) = \langle (e^t - 1)/t \,|\, (x)_n \rangle$$

the *Bernoulli numbers of the second kind.*

In order to expand $b_n(x)$ in powers of x^n, we observe that for $k \geq 1$

$$\frac{1}{k!}\langle t^k \mid b_n(x)\rangle = \frac{1}{k!}\left\langle t^k \left| \frac{e^t - 1}{t}(x)_n \right. \right\rangle$$

$$= \frac{1}{k!}\langle t^{k-1} \mid (e^t - 1)(x)_n\rangle$$

$$= \frac{n}{k!}\langle t^{k-1} \mid (x)_{n-1}\rangle$$

$$= \frac{n}{k}s(n-1, k-1), \qquad (4.3.7)$$

where $s(n, k)$ are the Stirling numbers of the first kind. Thus

$$b_n(x) = \sum_{k=0}^{n} \frac{1}{k!}\langle t^k \mid b_n(x)\rangle x^k = b_n(0) + \sum_{k=1}^{n} \frac{n}{k}s(n-1, k-1)x^k$$

(cf. the corresponding formula for Bernoulli polynomials in the subsection on definitions of Section 2.2).

Next we compute $\int_0^1 b_n(x)\,dx$ by using the formula for integration by parts, which is easily seen to be

$$\frac{e^t - 1}{t} = e^t - t\left(\frac{e^t - 1}{t}\right)'.$$

Then

$$\int_0^1 b_n(x)\,dx = \left\langle \frac{e^t - 1}{t} \middle| b_n(x) \right\rangle$$

$$= \left\langle e^t - t\left(\frac{e^t - 1}{t}\right)' \middle| b_n(x) \right\rangle$$

$$= b_n(1) - \left\langle \left(\frac{e^t - 1}{t}\right)' \middle| n(x)_{n-1} \right\rangle$$

$$= b_n(1) - n\left\langle \frac{e^t - 1}{t} \middle| x(x)_{n-1} \right\rangle$$

$$= b_n(1) - n\left\langle \frac{e^t - 1}{t} \middle| (x)_n + (n-1)(x)_{n-1} \right\rangle$$

$$= b_n(1) - nb_n(0) - n(n-1)b_{n-1}(0)$$

$$= (1 - n)b_n(0) + n(2 - n)b_{n-1}(0). \qquad (4.3.8)$$

GENERATING FUNCTION AND SHEFFER IDENTITY

The generating function for $b_n(x)$ is

$$\sum_{k=0}^{\infty} \frac{b_k(x)}{k!} t^k = \frac{t}{\log(1+t)} (1+t)^x.$$

If $x = 0$, we have the expansion

$$\sum_{k=0}^{\infty} \frac{b_k(0)}{k!} t^k = \frac{t}{\log(1+t)}.$$

The Sheffer identity is

$$b_n(x+y) = \sum_{k=0}^{n} \binom{n}{k} b_k(y)(x)_{n-k}.$$

Taking $y = 0$ gives

$$b_n(x) = \sum_{k=0}^{n} \binom{n}{k} b_k(0)(x)_{n-k}.$$

Applying t to this yields

$$\langle t \mid b_n(x) \rangle = \sum_{k=0}^{n} \binom{n}{k} b_k(0) \langle t \mid (x)_{n-k} \rangle$$

or, since for $n \geq 1$

$$\langle t \mid b_n(x) \rangle = \langle t^0 \mid n(x)_{n-1} \rangle = \delta_{n,1}$$

and

$$\langle t \mid (x)_{n-k} \rangle = (-1)^{n-k-1}(n-k-1)!,$$

we have

$$\delta_{n,1} = \sum_{k=0}^{n-1} \frac{(-1)^k}{k!(n-k)} b_k(0).$$

From this one can compute the numbers $b_n(0)$.

EXPANSIONS

The expansion theorem is

$$h(t) = \sum_{k=0}^{\infty} \frac{\langle h(t) \mid b_k(x) \rangle}{k!} \frac{t}{e^t - 1} (e^t - 1)^k.$$

Taking $h(t) = 1$ and multiplying by $(e^t - 1)/t$, we get

$$\frac{e^t - 1}{t} = \sum_{k=0}^{\infty} \frac{b_k(0)}{k!} (e^t - 1)^k.$$

This is Gregory's formula, encountered in the subsection on expansions in Section 1.2, where we pointed out that Gregory discovered this formula in 1670 and that it was reportedly the first formula for numerical integration.

Applying Gregory's formula to $b_n(x)$ gives another expression for $\int_0^1 b_n(x)\,dx$,

$$\int_0^1 b_n(x)\,dx = \left\langle \frac{e^t - 1}{t} \,\middle|\, b_n(x) \right\rangle$$

$$= \sum_{k=0}^n \frac{b_k(0)}{k!} \langle (e^t - 1)^k \mid b_n(x) \rangle$$

$$= \sum_{k=0}^n \binom{n}{k} b_k(0) b_{n-k}(0).$$

The polynomial expansion theorem is

$$p(x) = \sum_{k \geq 0} \frac{1}{k!} \left\langle \frac{t}{e^t - 1}(e^t - 1)^k \,\middle|\, p(x) \right\rangle b_k(x).$$

If we replace $p(x)$ by $[(e^t - 1)/t]\,p(x)$, we obtain a formula for the integral of a polynomial in terms of its differences at 0,

$$\int_x^{x+1} p(u)\,du = \frac{e^t - 1}{t} p(x) = \sum_{k \geq 0} \frac{1}{k!} \langle (e^t - 1)^k \mid p(x) \rangle b_k(x).$$

Let us next expand $b_n(x)$ in terms of the Bernoulli polynomials (of the first kind) $B_n(x)$. Using (4.3.7) and (4.3.8), we have

$$b_n(x) = \sum_{k=0}^n \frac{1}{k!} \left\langle \frac{e^t - 1}{t} t^k \,\middle|\, b_n(x) \right\rangle B_k(x)$$

$$= \left\langle \frac{e^t - 1}{t} \,\middle|\, b_n(x) \right\rangle + \langle e^t - 1 \mid b_n(x) \rangle B_1(x)$$

$$+ \sum_{k=2}^n \frac{1}{k!} \langle (e^t - 1) t^{k-1} \mid b_n(x) \rangle B_k(x)$$

$$= (1 - n) b_n(0) + n(2 - n) b_{n-1}(0) + n b_{n-1}(0) B_1(x)$$

$$+ \sum_{k=2}^n \frac{n}{k!} \langle t^{k-1} \mid b_{n-1}(x) \rangle B_k(x)$$

$$= (1 - n) b_n(0) + n(2 - n) b_{n-1}(0) + n b_{n-1}(0) B_1(x)$$

$$+ \sum_{k=2}^n \frac{n(n - 1)}{k(k - 1)} s(n - 2, k - 2) B_k(x).$$

We may expand $B_n(x)$ in terms of $b_n(x)$ by

$$B_n(x) = \sum_{k=0}^{n} \frac{1}{k!} \left\langle \frac{t}{e^t - 1}(e^t - 1)^k \,\middle|\, B_n(x) \right\rangle b_k(x).$$

In order to evaluate $\langle t/(e^t - 1) \,|\, B_n(x)\rangle$ we require the analog of integration by parts for the operator $t/(e^t - 1)$, which is

$$\frac{t}{e^t - 1} = 1 - t - (e^t - 1)\left(\frac{t}{e^t - 1}\right)'.$$

Then

$$
\begin{aligned}
\left\langle \frac{t}{e^t - 1} \,\middle|\, B_n(x) \right\rangle &= \left\langle 1 - t - (e^t - 1)\left(\frac{t}{e^t - 1}\right)' \,\middle|\, B_n(x) \right\rangle \\
&= \langle 1 - t \,|\, B_n(x)\rangle - \left\langle \left(\frac{t}{e^t - 1}\right)' \,\middle|\, (e^t - 1)B_n(x) \right\rangle \\
&= B_n(0) - nB_{n-1}(0) - \left\langle \left(\frac{t}{e^t - 1}\right)' \,\middle|\, nx^{n-1} \right\rangle \\
&= B_n(0) - nB_{n-1}(0) - \left\langle \frac{t}{e^t - 1} \,\middle|\, nx^n \right\rangle \\
&= B_n(0) - nB_{n-1}(0) - nB_n(0) \\
&= (1 - n)B_n(0) - nB_{n-1}(0).
\end{aligned}
$$

Hence

$$
\begin{aligned}
B_n(x) &= (1 - n)B_n(0) - nB_{n-1}(0) + \langle t \,|\, B_n(x)\rangle b_1(x) \\
&\quad + \sum_{k=2}^{n} \frac{1}{k!} \langle t(e^t - 1)^{k-1} \,|\, B_n(x)\rangle b_k(x) \\
&= (1 - n)B_n(0) - nB_{n-1}(0) + nB_{n-1}(0)b_1(x) \\
&\quad + \sum_{k=2}^{n} \frac{n}{k!} \langle (e^t - 1)^{k-2} \,|\, (e^t - 1)B_{n-1}(x)\rangle b_k(x) \\
&= (1 - n)B_n(0) - nB_{n-1}(0) + nB_{n-1}(0)b_1(x) \\
&\quad + \sum_{k=2}^{n} \frac{n(n-1)}{k!} \langle (e^t - 1)^{k-2} \,|\, x^{n-2}\rangle b_k(x) \\
&= (1 - n)B_n(0) - nB_{n-1}(0) + nB_{n-1}(0)b_1(x) \\
&\quad + \sum_{k=2}^{n} \frac{n(n-1)}{k(k-1)} S(n - 2, k - 2)b_k(x).
\end{aligned}
$$

Symmetry in the Bernoulli polynomials of the second kind is easily detected by observing that, according to Proposition 2.1.11, the sequence $b_n(-x)$ is Sheffer for

$$\left(-t/(e^t - 1), e^{-t} - 1\right).$$

Now

$$\langle (e^{-t} - 1)^k \,|\, (-1)^n x^{(n)} \rangle = (-1)^{n-k} \langle (1 - e^{-t})^k \,|\, x^{(n)} \rangle$$

$$= (-1)^{n-k} n!\, \delta_{n,k} = n!\, \delta_{n,k},$$

and so $(-1)^n x^{(n)}$ is associated to $e^{-t} - 1$. Thus

$$b_n(-x) = (-1)^n \frac{e^{-t} - 1}{-t} x^{(n)}$$

$$= (-1)^n e^{-t} \frac{e^t - 1}{t} (x + n - 1)_n$$

$$= (-1)^n e^{-t} b_n(x + n - 1)$$

$$= (-1)^n b_n(x + n - 2).$$

Replacing x by $1 - \tfrac{1}{2}n + x$, we get

$$b_n(\tfrac{1}{2}n - 1 - x) = (-1)^n b_n(\tfrac{1}{2}n - 1 + x).$$

3.3. The Poisson–Charlier Polynomials

The Poisson–Charlier polynomials $c_n(x;a)$ form the Sheffer sequence for

$$g(t) = e^{a(e^t - 1)},$$

$$f(t) = a(e^t - 1)$$

for $a \neq 0$. These polynomials are discussed by Jordan [1, p. 473], Erdelyi [1, Vol. 2, p. 226], and Szegö [1, p. 34]. They derive their importance from the fact that, for $a > 0$, they are orthogonal with respect to the Poisson distribution; that is,

$$\sum_{k=0}^{\infty} j(k) c_n(k;a) c_m(k;a) = a^{-n} n!\, \delta_{n,m},$$

where $j(k)$ is the Poisson density

$$j(k) = (a^k/k!) e^{-a}$$

for $k = 0, 1, 2, \ldots$.

Since

$$\langle a^k(e^t - 1)^k | a^{-n}(x)_n \rangle = a^{k-n}\langle (e^t - 1)^k | (x)_n \rangle$$
$$= n!\,\delta_{n,k},$$

we see that the associated sequence for $f(t)$ is $p_n(x) = a^{-n}(x)_n$. Thus from Theorem 2.3.6 we obtain

$$c_n(x;a) = g(t)^{-1} p_n(x)$$

$$= a^{-n} e^{-a(e^t - 1)}(x)_n$$

$$= a^{-n} \sum_{k=0}^{n} \frac{(-a)^k}{k!}(e^t - 1)^k (x)_n$$

$$= a^{-n} \sum_{k=0}^{n} \binom{n}{k}(-a)^k (x)_{n-k}$$

$$= \sum_{k=0}^{n} \binom{n}{k}(-1)^{n-k} a^{-k} (x)_k.$$

Recalling that $\langle t^j | (x)_k \rangle = j!\,s(k,j)$, we get the explicit expression

$$c_n(x;a) = \sum_{j=0}^{n} \frac{\langle t^j | c_n(x;a) \rangle}{j!} x^j$$

$$= \sum_{j=0}^{n} \sum_{k=0}^{n} \binom{n}{k}(-1)^{n-k} a^{-k} \left\langle \frac{t^j}{j!} \bigg| (x)_k \right\rangle x^j$$

$$= \sum_{j=0}^{n} \left[\sum_{k=0}^{n} \binom{n}{k}(-1)^{n-k} a^{-k} s(k,j) \right] x^j.$$

It is interesting to note that

$$a^{-n} L_n^{(x-n)}(a) = a^{-n} \sum_{k=0}^{n} \binom{x}{n-k} \frac{n!}{k!}(-a)^k$$

$$= a^{-n} \sum_{k=0}^{n} \binom{n}{k}(-a)^{n-k} (x)_k$$

$$= c_n(x;a),$$

where $L_n^{(\alpha)}(x)$ are the Laguerre polynomials.

GENERATING FUNCTION AND SHEFFER IDENTITY

The generating function for the Poisson–Charlier polynomials is

$$\sum_{k=0}^{\infty} \frac{c_k(x;a)}{k!} t^k = e^{-t}\left(1 + \frac{t}{a}\right)^x.$$

The Sheffer identity is

$$c_n(x + y; a) = \sum_{k=0}^{n} \binom{n}{k} a^{k-n} c_k(y; a)(x)_{n-k}.$$

RECURRENCE RELATIONS AND OPERATIONAL FORMULAS

Theorem 2.3.7 is

$$a(e^t - 1)c_n(x; a) = nc_{n-1}(x; a),$$

which is the recurrence

$$ac_n(x + 1; a) - ac_n(x; a) - nc_{n-1}(x; a) = 0. \tag{4.3.9}$$

The recurrence formula is

$$
\begin{aligned}
c_{n+1}(x; a) &= \left(x - \frac{g'(t)}{g(t)}\right)\frac{1}{f'(t)}c_n(x; a) \\
&= (x - ae^t)\frac{1}{ae^t}c_n(x; a) \\
&= a^{-1}xc_n(x - 1; a) - c_n(x; a)
\end{aligned}
$$

or

$$ac_{n+1}(x; a) + ac_n(x; a) - xc_n(x - 1; a) = 0. \tag{4.3.10}$$

If we solve for $c_{n-1}(x; a)$ in (4.3.9) and substitute into (4.3.10), where n is replaced by $n - 1$, we get

$$ac_n(x + 1; a) + (n - a - x)c_n(x; a) + xc_n(x - 1; a) = 0.$$

EXPANSIONS

Referring to Theorem 2.3.11, we have

$$
\begin{aligned}
\langle g'(t)\,|\,c_{n-k}(x; a)\rangle &= \langle ae^t e^{a(e^t - 1)}\,|\,c_{n-k}(x; a)\rangle \\
&= \langle ae^t\,|\,a^{k-n}(x)_{n-k}\rangle \\
&= a^{k-n+1}(1)_{n-k} \\
&= a^{k-n+1}(\delta_{n,k} + \delta_{n-1,k})
\end{aligned}
$$

and

$$
\begin{aligned}
\langle g(t)f'(t)\,|\,c_{n-k+1}(x; a)\rangle &= \langle ae^t e^{a(e^t - 1)}\,|\,c_{n-k+1}(x; a)\rangle \\
&= a^{k-n}(\delta_{n+1,k} + \delta_{n,k}).
\end{aligned}
$$

Thus

$$xc_n(x;a) = \sum_{k=0}^{n+1}\left[\binom{n}{k}a^{k-n+1}(\delta_{n,k} + \delta_{n-1,k})\right.$$

$$\left. + \binom{n}{k-1}a^{k-n}(\delta_{n+1,k} + \delta_{n,k})\right]c_k(x;a)$$

$$= ac_{n+1}(x;a) + (a + n)c_n(x;a) + nc_{n-1}(x;a)$$

or

$$ac_{n+1}(x;a) + (a + n - x)c_n(x;a) + nc_{n-1}(x;a) = 0.$$

According to Theorem 2.3.12,

$$c_n'(x;a) = \sum_{k=0}^{n-1}\binom{n}{k}\langle t \mid a^{k-n}(x)_{n-k}\rangle c_k(x;a)$$

$$= \sum_{k=0}^{n-1}\binom{n}{k}a^{k-n}(-1)^{n-k-1}(n - k - 1)!\, c_k(x;a).$$

MISCELLANEOUS RESULTS

An interesting feature of the Poisson–Charlier sequence is its inverse under umbral composition. Theorem 3.5.5 implies that this inverse $s_n(x)$ is Sheffer for

$$(g(\bar{f}(t))^{-1}, \bar{f}(t)) = \left(e^{-t}, \log\left(1 + \frac{t}{a}\right)\right).$$

Proposition 2.1.11 tells us that the associated sequence for $\log(1 + t/a)$ is $\phi_n(ax)$, where $\phi_n(x)$ are the exponential polynomials. Thus

$$s_n(x) = e^t\phi_n(ax) = \phi_n(a(x + 1)),$$

and so

$$c_n(\phi(x)) = ((x - a)/a)^n.$$

Recalling a formula in the subsection on miscellaneous results of Section 1.3,

$$p(\phi(x)) = e^{-x}\sum_{k=0}^{\infty}\frac{p(k)}{k!}x^k,$$

which is valid for any polynomial $p(x)$, and letting $p(x) = c_n(x;a)$, we have

$$\sum_{k=0}^{\infty}c_n(k;a)\frac{x^k}{k!}e^{-x} = \left(\frac{x - a}{a}\right)^n.$$

3.4. The Actuarial Polynomials

DEFINITIONS

The polynomials $a_n^{(\beta)}(x)$, Sheffer for

$$g(t) = (1 - t)^{-\beta},$$

$$f(t) = \log(1 - t),$$

are known as the actuarial polynomials, since they were introduced in connection with problems of actuarial mathematics. They are discussed briefly in Erdelyi [1, Vol. 3, p. 254] and Boas and Buck [1, p. 42].

Since the associated sequence for $f(t) = \log(1 - t)$ is $p_n(x) = \phi_n(-x)$, where $\phi_n(x)$ are the exponential polynomials, we have

$$a_n^{(\beta)}(x) = (1 - t)^{\beta} \phi_n(-x)$$

$$= \sum_{k=0}^{\infty} \binom{\beta}{k}(-1)^k t^k \phi_n(-x) = \sum_{k=0}^{n} \binom{\beta}{k} \phi_n^{(k)}(-x). \qquad (4.3.11)$$

Also,

$$(1 - t)^{\lambda} a_n^{(\beta)}(x) = a_n^{(\beta + \lambda)}(x). \qquad (4.3.12)$$

Expanding $\phi_n^{(k)}(-x)$ and substituting into (4.3.11), we get the explicit expression

$$a_n^{(\beta)}(x) = \sum_{k=0}^{n} \binom{\beta}{k} \sum_{j=k}^{n} S(n, j)(j)_k(-x)^{j-k}$$

$$= \sum_{l=0}^{n} \left[\sum_{k=0}^{n-1} \binom{l+k}{k} (\beta)_k S(n, l+k) \right] (-x)^l.$$

GENERATING FUNCTION AND SHEFFER IDENTITY

The generating function for $a_n^{(\beta)}(x)$ is

$$\sum_{k=0}^{\infty} \frac{a_k^{(\beta)}(x)}{k!} t^k = \exp[\beta t + x(1 - e^t)].$$

The Sheffer identity is

$$a_n^{(\beta)}(x + y) = \sum_{k=0}^{n} \binom{n}{k} a_k^{(\beta)}(y) \phi_{n-k}(-x),$$

and applying $(1 - t)^{\lambda}$ gives

$$a_n^{(\beta + \lambda)}(x + y) = \sum_{k=0}^{n} \binom{n}{k} a_k^{(\beta)}(y) a_{n-k}^{(\lambda)}(x).$$

RECURRENCE RELATIONS AND OPERATIONAL FORMULAS

Letting $\lambda = 1$ in (4.3.12), we obtain the formula

$$a_n^{(\beta+1)}(x) = a_n^{(\beta)}(x) - ta_n^{(\beta)}(x). \qquad (4.3.13)$$

The recurrence formula is

$$a_{n+1}^{(\beta)}(x) = \left(x - \frac{g'(t)}{g(t)}\right)\frac{1}{f'(t)}a_n^{(\beta)}(x)$$
$$= -(x - \beta(1-t)^{-1})(1-t)a_n^{(\beta)}(x)$$
$$= (xt - x + \beta)a_n^{(\beta)}(x),$$

and so

$$a_n^{(\beta)}(x) = (xt - x + \beta)^n 1$$

(cf. the corresponding formula for Laguerre polynomials in the subsection on recurrence relations and operational formulas of Section 3.1). Using (4.3.12), we have

$$a_{n+1}^{(\beta)}(x) = -(x - \beta(1-t)^{-1})(1-t)a_n^{(\beta)}(x)$$
$$= -xa_n^{(\beta+1)}(x) + \beta a_n^{(\beta)}(x).$$

EXPANSIONS

For use in Theorem 2.3.11 we compute

$$\langle g'(t) | a_{n-k}^{(\beta)}(x)\rangle = \langle \beta(1-t)^{-\beta-1} | a_{n-k}^{(\beta)}(x)\rangle$$
$$= \beta a_{n-k}^{(-1)}(0)$$

and

$$\langle g(t)f'(t) | a_{n-k+1}^{(\beta)}(x)\rangle = \langle -(1-t)^{-\beta-1} | a_{n-k+1}^{(\beta)}(x)\rangle$$
$$= -a_{n-k+1}^{(-1)}(0).$$

Therefore,

$$xa_n^{(\beta)}(x) = \sum_{k=0}^{n+1}\left[\binom{n}{k}\beta a_{n-k}^{(-1)}(0) - \binom{n}{k-1}a_{n-k+1}^{(-1)}(0)\right]a_k^{(\beta)}(x).$$

Theorem 2.3.12 is

$$ta_n^{(\beta)}(x) = \sum_{k=0}^{n-1}\binom{n}{k}\langle 1 - e^t | x^{n-k}\rangle a_k^{(\beta)}(x) = -\sum_{k=0}^{n-1}\binom{n}{k}a_k^{(\beta)}(x).$$

Combining the last result with (4.3.13) gives the recurrence

$$a_n^{(\beta+1)}(x) = \sum_{k=0}^{n}\binom{n}{k}a_k^{(\beta)}(x).$$

MISCELLANEOUS RESULTS

Let us conclude with an umbral composition. The sequence $a_n^{(\beta)}(-x)$ is Sheffer for $\left((1+t)^{-\beta}, \log(1+t)\right)$ according to Proposition 2.1.11. By Theorem 3.5.5 the inverse, under umbral composition, of $a_n^{(\beta)}(-x)$ is the Sheffer sequence $s_n(x)$ for

$$(e^{\beta t}, e^t - 1),$$

and so

$$s_n(x) = e^{-\beta t}(x)_n.$$

Thus

$$s_n\big(a^{(\beta)}(-x)\big) = x^n = e^{-x} \sum_{k=0}^{\infty} \frac{(k)_n}{k!} x^k$$

$$= e^{-x} \sum_{k=0}^{\infty} \frac{\langle e^{kt} \mid e^{\beta t} s_n(x) \rangle}{k!} x^k,$$

and so for any polynomial $p(x)$ we have

$$p\big(a^{(\beta)}(-x)\big) = e^{-x} \sum_{k=0}^{\infty} \frac{\langle e^{kt} \mid e^{\beta t} p(x) \rangle}{k!} x^k$$

$$= e^{-x} \sum_{k=0}^{\infty} \frac{p(k+\beta)}{k!} x^k.$$

Taking $p(x) = x^n$, we have shown (Whittaker and Watson [1, p. 336]) that

$$a_n^{(\beta)}(-x) = e^{-x} \sum_{k=0}^{\infty} \frac{(k+\beta)^n}{k!} x^k.$$

4. OTHER SHEFFER SEQUENCES

We conclude this chapter by mentioning some additional examples of Sheffer sequences.

4.1. The Meixner Polynomials of the First Kind

These polynomials form the Sheffer sequence for

$$g(t) = \left(\frac{1-c}{1-ce^t} \right)^{\beta},$$

$$f(t) = \frac{1-e^t}{c^{-1} - e^t}.$$

for $c \neq 1$. The generating function is

$$\sum_{k=0}^{\infty} \frac{m_k(x;\beta,c)}{k!} t^k = \left(1 - \frac{t}{c}\right)^x (1 - t)^{-x-\beta}.$$

Meixner has shown that this sequence of polynomials satisfies the familiar three-term recurrence relation

$$s_{n+1}(x) = (x - b_n)s_n(x) - d_n s_{n-1}(x),$$

$s_0(x) = 1$, which characterizes orthogonal polynomial sequences (Chihara [1, p. 175], Erdelyi [1, Vol. 2, p. 225]). We shall discuss this recurrence relation in the next chapter. Incidentally, the Krawtchouk polynomials, which are orthogonal with respect to the binomial distribution, are a special case of the Meixner polynomials of the first kind (Erdelyi [1, Vol. 2, p. 224], Szegö [1, p. 35]).

4.2. The Meixner Polynomials of the Second Kind

These polynomials are Sheffer for

$$g(t) = [(1 + \delta f(t))^2 + f(t)^2]^{\eta/2},$$

$$f(t) = \tan \frac{t}{1 + \delta t},$$

and so the generating function is

$$\sum_{k=0}^{\infty} \frac{M_k(x;\delta,\eta)}{k!} t^k = [(1 + \delta t)^2 + t^2]^{-\eta/2} \exp\left[x \tan^{-1}\left(\frac{t}{1 + \delta t}\right)\right].$$

Like the Meixner polynomials of the first kind, these polynomials form an orthogonal sequence (Chihara [1, p. 179]). We mention them again in this connection in the next chapter.

4.3. The Pidduck Polynomials

These polynomials are Sheffer for

$$g(t) = \frac{2}{e^t - 1},$$

$$f(t) = \frac{e^t - 1}{e^t + 1},$$

and the generating function is

$$\sum_{k=0}^{\infty} \frac{P_k(x)}{k!} t^k = (1 - t)^{-1} \left(\frac{1 + t}{1 - t} \right)^x.$$

We recall that the associated sequence for $f(t)$ is formed by the Mittag-Leffler polynomials $M_n(x)$, and so

$$P_n(x) = \tfrac{1}{2}(e^t + 1)M_n(x).$$

Bateman [1], Erdelyi [1, Vol. 3, p. 248], and Boas and Buck [1, p. 38] discuss the Pidduck polynomials.

4.4. The Narumi Polynomials

The Narumi polynomials form the Sheffer sequence for

$$g(t) = \left(\frac{e^t - 1}{t} \right)^a, \qquad f(t) = e^t - 1.$$

The generating function is

$$\sum_{k=0}^{\infty} \frac{s_k(x)}{k!} t^k = \left(\frac{t}{\log(1 + t)} \right)^a (1 + t)^x.$$

The Narumi polynomials are mentioned in Boas and Buck [1, p. 37] and Erdelyi [1, Vol. 3, p. 258, Eq. (3)].

4.5. The Boole Polynomials

These form the Sheffer sequence for

$$g(t) = 1 + e^{\lambda t},$$
$$f(t) = e^t - 1.$$

Thus the generating function is

$$\sum_{k=0}^{\infty} \frac{s_k(x)}{k!} t^k = [1 + (1 + t)^\lambda]^{-1}(1 + t)^x.$$

Jordan [1] and Boas and Buck [1, p. 37] discuss these polynomials. Actually, Jordan gives a detailed discussion of the polynomials $n! \, r_n(x)$, where $r_n(x)$ is Sheffer for

$$g(t) = \frac{1 + e^t}{2}, \qquad f(t) = e^t - 1.$$

4.6. The Peters Polynomials

These are a generalization of the Boole polynomials, being Sheffer for

$$g(t) = (1 - e^{\lambda t})^\mu,$$

$$f(t) = e^t - 1.$$

Thus

$$\sum_{k=0}^{\infty} \frac{s_k(x)}{k!} t^k = [1 + (1 + t)^\lambda]^{-\mu} (1 + t)^x.$$

They are mentioned in Boas and Buck [1, p. 37].

4.7. The Squared Hermite Polynomials

According to Erdelyi [1, Vol. 3, p. 250], if $H_n(x)$ are the Hermite polynomials (by our definition, not that of Erdelyi), then

$$\sum_{k=0}^{\infty} \frac{[H_k(x^{1/2})]^2}{k!} t^k = (1 - t^2)^{-1/2} e^{x[t/(1 + t)]}.$$

Thus the squared Hermite polynomials $[H_n(x^{1/2})]^2$ form the Sheffer sequence for

$$g(t) = (1 - 2t)^{1/2}/(1 - t),$$

$$f(t) = t/(1 - t).$$

See also Boas and Buck [1, p. 41].

4.8. The Stirling Polynomials

The Stirling polynomials $S_n(x)$ form the Sheffer sequence for $(g(t), f(t))$, where

$$g(t) = e^{-t},$$

$$\bar{f}(t) = \log\left(\frac{t}{1 - e^{-t}}\right).$$

The generating function is therefore

$$\sum_{k=0}^{\infty} \frac{S_k(x)}{k!} t^k = \left(\frac{t}{1 - e^{-t}}\right)^{x+1}.$$

(cf. Erdelyi [1, Vol. 3, p. 257]).

The Stirling polynomials may be connected to the Stirling numbers as follows. By Theorem 3.5.6 and Proposition 2.1.11 we have

$$S_n(m) = \langle e^{mt} \mid S_n(x) \rangle = \langle g(\bar{f}(t))^{-1} e^{m\bar{f}(t)} \mid x^n \rangle$$

$$= \left\langle \left(\frac{t}{1 - e^{-t}} \right)^{m+1} \middle| x^n \right\rangle$$

$$= (-1)^n \left\langle \left(\frac{t}{e^t - 1} \right)^{m+1} \middle| x^n \right\rangle.$$

Now, if m is an integer and $m \geq n$, then

$$S_n(m) = (-1)^n \left\langle \left(\frac{t}{e^t - 1} \right)^{m+1} \middle| x^n \right\rangle = (-1)^n B_n^{(m+1)}(0)$$

$$= \frac{(-1)^n}{\binom{m}{n}} s(m+1, m-n+1),$$

where $B_n^{(m+1)}(x)$ are the Bernoulli polynomials, $s(m+1, m-n+1)$ are the Stirling numbers of the first kind, and the last equality follows from a result obtained in the subsection on miscellaneous results of Section 2.2.

If m is a negative integer, then

$$S_n(m) = (-1)^n \left\langle \left(\frac{e^t - 1}{t} \right)^{-m-1} \middle| x^n \right\rangle$$

$$= (-1)^n \left\langle \left(\frac{e^t - 1}{t} \right)^{-m-1} \middle| t^{-m-1} \frac{1}{(n-m-1)_{-m-1}} x^{n-m-1} \right\rangle$$

$$= (-1)^n \left\langle (e^t - 1)^{-m-1} \middle| \frac{1}{(n-m-1)_{-m-1}} x^{n-m-1} \right\rangle$$

$$= \frac{(-1)^n n!}{(n-m-1)!} \langle t^{-m-1} \mid \phi_{n-m-1}(x) \rangle$$

$$= \frac{(-1)^n n!}{(n-m-1)!} S(n-m-1, -m-1),$$

where $S(n-m-1, -m-1)$ are the Stirling numbers of the second kind.

4.9. The Mahler Polynomials

Mahler introduced the polynomials $s_n(x)$, associated to $f(t)$, where

$$\bar{f}(t) = 1 + t - e^t,$$

for the investigation of the zeros of the incomplete gamma function (Erdelyi [1, Vol. 3, p. 254]). The generating function is

$$\sum_{k=0}^{\infty} \frac{s_k(x)}{k!} t^k = e^{x(1 + t - e^t)}.$$

4.10. Polynomials Related to the Hermite Polynomials

The associated sequence for $f(t)$, where

$$\bar{f}(t) = t - \tfrac{1}{2}t^2 - \log(1 + t) = -\tfrac{1}{3}t^3 + \tfrac{1}{4}t^4 - \cdots$$

is related to the Hermite polynomials and was introduced for the study of the asymptotic properties of the Hermite polynomials (Erdelyi [1, Vol. 3, p. 256]). The generating function is

$$\sum_{k=0}^{\infty} \frac{s_k(x)}{k!} t^k = (1 + t)^{-x} e^{x(t - t^2/2)}.$$

4.11. Polynomials Related to Hyperbolic Differential Equations

The Sheffer sequence for

$$g(t) = (1 + t)^{-c},$$

$$f(t) = 1 - (1 + t)^{-2}$$

has been applied to the theory of hyperbolic differential equations (Erdelyi [1, Vol. 3, pp. 257–258]). The generating function is

$$\sum_{k=0}^{\infty} \frac{s_k(x)}{k!} t^k = (1 + t)^{-x} e^{x(t - t^2/2)}.$$

4.12. The Mott Polynomials

Mott has considered the associated sequence for

$$f(t) = -2t/(1 - t^2)$$

in connection with problems in the theory of electrons (Erdelyi [1, Vol. 3, p. 251]). The generating function is

$$\sum_{k=0}^{\infty} \frac{s_k(x)}{k!} t^k = \exp\left(x \left[\frac{(1 - t^2)^{1/2} - 1}{t} \right] \right).$$

CHAPTER 5

TOPICS

In this chapter we discuss some topics to which umbral methods can be fruitfully applied. We shall consider only enough examples in each case to give an indication of the success of the method. The sections of this chapter may be read independently of each other.

1. THE CONNECTION CONSTANTS PROBLEM AND DUPLICATION FORMULAS

The connection constants problem consists in determining the connection constants $c_{n,k}$ in the expression

$$s_n(x) = \sum_{k=0}^{n} c_{n,k} r_k(x), \qquad (5.1.1)$$

where $s_n(x)$ and $r_n(x)$ are sequences of polynomials. Umbral methods give an easy solution to this problem when the sequences involved are Sheffer.

Let $s_n(x)$ be Sheffer for $(g(t), f(t))$ and let $r_n(x)$ be Sheffer for $(h(t), l(t))$. In (5.1.1) we recognize an umbral composition

$$s_n(x) = \lambda_{h(t), l(t)} \sum_{k=0}^{n} c_{n,k} x^k,$$

and so from Theorem 3.5.2 we deduce that

$$t_n(x) = \sum_{k=0}^{n} c_{n,k} x^k$$

$$= \lambda_{h(t),\, l(t)}^{-1} s_n(x)$$

$$= \lambda_{h(t),\, l(t)}^{-1} \lambda_{f(t),\, g(t)} x^n$$

$$= \lambda_{h(\bar{l}(t))\bar{l}(t)} \lambda_{f(t),\, g(t)} x^n$$

$$= \lambda_{g(\bar{l}(t))h(\bar{l}(t))^{-1},\, f(\bar{l}(t))} x^n.$$

Thus the sequence

$$t_n(x) = \sum_{k=0}^{n} c_{n,k} x^k$$

is Sheffer for the pair

$$\left(\frac{g(\bar{l}(t))}{h(\bar{l}(t))},\, f(\bar{l}(t)) \right).$$

This is one form of our solution. To get an explicit expression for the connection constants we use (3.5.2),

$$c_{n,k} = \frac{1}{k!} \langle t^k \mid t_n(x) \rangle$$

$$= \frac{1}{k!} \langle \lambda^*_{g(\bar{l}(t))h(\bar{l}(t))^{-1},\, f(\bar{l}(t))} t^k \mid x^n \rangle$$

$$= \frac{1}{k!} \left\langle \frac{h(\bar{f}(t))}{g(\bar{f}(t))} l(\bar{f}(t))^k \,\middle|\, x^n \right\rangle.$$

A duplication formula is a formula of the form

$$r_n(ax) = \sum_{k=0}^{n} c_{n,k} r_k(x),$$

where a is a nonzero constant.

When $r_n(x)$ is a Sheffer sequence, this is a special case of the connection constants problem, for if $r_n(x)$ is Sheffer for $(h(t), l(t))$, then, according to Proposition 2.1.11,

$$\langle h(a^{-1}t) l(a^{-1}t)^k \mid r_n(ax) \rangle = \langle h(t) l(t)^k \mid r_n(x) \rangle$$

$$= n! \, \delta_{n,k}$$

and so $r_n(ax)$ is Sheffer for $(h(a^{-1}t), l(a^{-1}t))$. Thus we see that

$$t_n(x) = \sum_{k=0}^{n} c_{n,k} x^k$$

is Sheffer for

$$\left(\frac{h(a^{-1}\bar{l}(t))}{h(\bar{l}(t))}, l(a^{-1}\bar{l}(t))\right)$$

and

$$c_{n,k} = \frac{1}{k!}\left\langle \frac{h(a\bar{l}(t))}{h(\bar{l}(t))} l(a\bar{l}(t))^k \,\Big|\, x^n \right\rangle.$$

Let us give some examples.

Example 1   Consider the formula

$$(x)_n = \sum_{k=0}^n c_{n,k} x^{(n)},$$

where $x^{(n)} = x(x+1)\cdots(x+n-1)$ is associated to the backward difference $1 - e^{-t}$. We have

$$g(t) = 1, \qquad f(t) = e^t - 1,$$
$$h(t) = 1, \qquad l(t) = 1 - e^{-t},$$

and so $t_n(x)$ is Sheffer for

$$\left(\frac{g(\bar{l}(t))}{h(\bar{l}(t))}, f(\bar{l}(t))\right) = \left(1, -\frac{t}{t-1}\right).$$

Thus

$$t_n(x) = L_n(-x),$$

where $L_n(x)$ are the Laguerre polynomials. In particular,

$$t_n(x) = \sum_{k=1}^n \binom{n-1}{k-1}\frac{n!}{k!} x^k$$

and

$$(x)_n = \sum_{k=1}^n \binom{n-1}{k-1}\frac{n!}{k!} x^{(n)}.$$

Example 2   Consider the formula

$$x^{[n]} = \sum_{k=0}^n c_{n,k}(x)_k.$$

The central factorials are Sheffer for

$$g(t) = 1, \qquad f(t) = e^{t/2} - e^{-t/2},$$

and the lower factorials are Sheffer for

$$h(t) = 1, \qquad l(t) = e^t - 1.$$

Thus the sequence $t_n(x)$ is Sheffer for

$$\left(1, \frac{t}{(1+t)^{1/2}}\right).$$

By the transfer formula

$$t_n(x) = x(1+t)^{n/2} x^{n-1}$$

$$= \sum_{k=0}^{n} \binom{n/2}{k} (n-1)_k x^{n-k}$$

$$= \sum_{k=0}^{n} \binom{n/2}{n-k} (n-1)_{n-k} x^k,$$

and so

$$x^{[n]} = \sum_{k=0}^{n} \binom{n/2}{n-k} (n-1)_{n-k} (x)_k.$$

Example 3   Consider next

$$L_n^{(\alpha)}(x) = \sum_{k=0}^{n} c_{n,k} \phi_k(-x),$$

where $L_n^{(\alpha)}(x)$ are the Laguerre polynomials and $\phi_n(x)$ are the exponential polynomials. Here we have

$$g(t) = (1-t)^{-\alpha-1}, \qquad f(t) = t/(t-1)$$

and

$$h(t) = 1, \qquad\qquad l(t) = \log(1-t).$$

Thus $t_n(x)$ is Sheffer for

$$(e^{-(\alpha+1)t}, 1 - e^{-t})$$

and so

$$t_n(x) = e^{(\alpha+1)t}(x)^{(n)}$$

$$= (x + \alpha + 1)^{(n)}$$

$$= \sum_{j=0}^{n} |s(n,j)| (x + \alpha + 1)^j \qquad \text{(Comtet [1, p. 213])}$$

$$= \sum_{k=0}^{n} \left[\sum_{j=k}^{n} |s(n,j)| \binom{j}{k} (\alpha+1)^{j-k} \right] x^k,$$

where $s(n,j)$ are the Stirling numbers of the first kind. Thus we have

$$L_n^{(\alpha)}(x) = \sum_{k=0}^{n} \left[\sum_{j=k}^{n} \binom{j}{k} |s(n,j)| (\alpha+1)^{j-k} \right] \phi_k(-x).$$

For $\alpha = -1$ we get

$$L_n(x) = \sum_{k=0}^{n} |s(n, k)| \, \phi_k(-x).$$

Example 4   Consider

$$H_n^{(v)}(x) = \sum_{k=0}^{n} c_{n,k} H_k^{(\lambda)}(x),$$

where $H_n^{(v)}(x)$ are the Hermite polynomials of variance v. Here we have

$$g(t) = e^{vt^2/2}, \qquad f(t) = t,$$
$$h(t) = e^{\lambda t^2/2}, \qquad l(t) = t.$$

Thus $t_n(x)$ is Sheffer for

$$(e^{(v-\lambda)t^2/2}, t)$$

and

$$
\begin{aligned}
t_n(x) &= e^{(\lambda - v)t^2/2} x^n \\
&= \sum_{j \geq 0} \left(\frac{\lambda - v}{2} \right)^j \frac{(n)_{2j}}{j!} x^{n-2j} \\
&= \sum_{\substack{k \geq 0 \\ n-k \text{ even}}} \left(\frac{\lambda - v}{2} \right)^{(n-k)/2} \frac{(n)_{n-k}}{(\frac{1}{2}(n-k))!} x^k,
\end{aligned}
$$

and so

$$H_n^{(v)}(x) = \sum_{\substack{k=0 \\ n-k \text{ even}}}^{n} \left(\frac{\lambda - v}{2} \right)^{(n-k)/2} \frac{(n)_{n-k}}{(\frac{1}{2}(n-k))!} H_k^{(\lambda)}(x).$$

Example 5   Let us consider

$$E_n(x) = \sum_{k=0}^{n} c_{n,k} B_k(x),$$

where $E_n(x)$ are the Euler polynomials and $B_n(x)$ are the Bernoulli polynomials. Here

$$g(t) = (e^t + 1)/2, \qquad f(t) = t,$$
$$h(t) = (e^t - 1)/t, \qquad l(t) = t,$$

and so $t_n(x)$ is Sheffer for

$$\left(\frac{t}{e^t - 1} \frac{e^t + 1}{2}, t \right).$$

Hence

$$
t_n(x) = \frac{e^t - 1}{t} \frac{2}{e^t + 1} x^n = \frac{e^t - 1}{t} E_n(x)
$$

$$
= \frac{e^t - 1}{t} \frac{1}{n + 1} t E_{n+1}(x) = \frac{1}{n + 1}(e^t - 1) E_{n+1}(x)
$$

$$
= \frac{1}{n + 1}(e^t + 1 - 2) E_{n+1}(x) = \frac{2}{n + 1}(x^{n+1} - E_{n+1}(x))
$$

$$
= \sum_{k=0}^{n} \binom{n+1}{k} \frac{-2}{n+1} E_{n+1-k}(0) x^k,
$$

and so

$$
E_n(x) = \sum_{k=0}^{n} \binom{n+1}{k} \frac{-2}{n+1} E_{n+1-k}(0) B_k(x).
$$

Example 6 Consider

$$
c_n(x; b) = \sum_{k=0}^{n} c_{n,k} c_k(x; a),
$$

where $c_n(x; a)$ are the Poisson–Charlier polynomials. In this case

$$
g(t) = e^{b(e^t - 1)}, \qquad f(t) = b(e^t - 1),
$$

$$
h(t) = e^{a(e^t - 1)}, \qquad l(t) = a(e^t - 1),
$$

and so $t_n(x)$ is Sheffer for

$$
\left(e^{(b/a - 1)t}, \frac{b}{a} t \right).
$$

Therefore

$$
t_n(x) = \left(\frac{a}{b} \right)^n e^{(1 - b/a)t} x^n
$$

$$
= \left(\frac{a}{b} \right)^n \left(x + 1 - \frac{b}{a} \right)^n
$$

$$
= \left(\frac{a}{b} \right)^n \sum_{k=0}^{n} \binom{n}{k} \left(1 - \frac{b}{a} \right)^{n-k} x^k
$$

and

$$
c_n(x; b) = \left(\frac{a}{b} \right)^n \sum_{k=0}^{n} \binom{n}{k} \left(1 - \frac{b}{a} \right)^{n-k} c_k(x; a).
$$

Example 7   Consider the duplication formula for Hermite polynomials,

$$H_n^{(v)}(ax) = \sum_{k=0}^{n} c_{n,k} H_n^{(v)}(x).$$

We have

$$h(t) = \exp(vt^2/2), \qquad l(t) = t,$$

and so

$$c_{n,k} = \frac{1}{k!} \langle \exp[(a^2 - 1)vt^2/2] a^k t^k \,|\, x^n \rangle$$

$$= \binom{n}{k} a^k \langle \exp[(a^2 - 1)vt^2/2] \,|\, x^{n-k} \rangle$$

$$= \begin{cases} \binom{n}{k}\left(\dfrac{v}{2}\right)^{(n-k)/2} (a^2 - 1)^{(n-k)/2} a^k \dfrac{1}{(\frac{1}{2}(n-k))!}, & n-k \quad \text{even}, \\ 0, & n-k \quad \text{odd}. \end{cases}$$

Thus

$$H_n^{(v)}(ax) = \sum_{\substack{k=0 \\ n-k \text{ even}}}^{n} \binom{n}{k}\left(\frac{v}{2}\right)^{(n-k)/2} (a^2 - 1)^{(n-k)/2} a^k \frac{1}{(\frac{1}{2}(n-k))!} H_k^{(v)}(x).$$

Example 8   Consider the duplication formula for Laguerre polynomials,

$$L_n^{(\alpha)}(ax) = \sum_{k=0}^{n} c_{n,k} L_n^{(\alpha)}(x).$$

We have

$$h(t) = (1 - t)^{-\alpha - 1}, \qquad l(t) = t/(t - 1),$$

and so

$$c_{n,k} = \frac{1}{k!} \left\langle (at - t + 1)^{-\alpha - 1} \left(\frac{at}{at - t + 1}\right)^k \,\middle|\, x^n \right\rangle$$

$$= \frac{a^k}{k!} \langle (at - t + 1)^{-\alpha - 1 - k} t^k \,|\, x^n \rangle$$

$$= \binom{n}{k} a^k \langle [(a - 1)t + 1]^{-\alpha - 1 - k} \,|\, x^{n-k} \rangle$$

$$= \binom{n}{k} a^k \sum_{j=0}^{\infty} \binom{-\alpha - 1 - k}{j} (a - 1)^j \langle t^j \,|\, x^{n-k} \rangle$$

$$= \binom{n}{k} a^k (a - 1)^{n-k} \binom{-\alpha - 1 - k}{n - k} = \binom{n + \alpha}{n - k} \frac{n!}{k!} a^k (1 - a)^{n-k}.$$

Thus we obtain

$$L_n^{(\alpha)}(ax) = \sum_{k=0}^{n} \binom{n+\alpha}{n-k} \frac{n!}{k!} a^k (1-a)^{n-k} L_k^{(\alpha)}(x).$$

Example 9   Let us obtain the coefficients in

$$M_n\left(\frac{x}{2}\right) = \sum_{k=0}^{n} c_{n,k} M_k(x),$$

where $M_n(x)$ are the Mittag-Leffler polynomials. Recall that $M_n(x)$ are associated to

$$l(t) = (e^t - 1)/(e^t + 1)$$

and that

$$\bar{l}(t) = \log((1+t)/(1-t)).$$

Thus the sequence $t_n(x)$ is associated to

$$l(2\bar{l}(t)) = \frac{((1+t)/(1-t))^2 - 1}{((1+t)/(1-t))^2 + 1}$$

$$= \frac{(1+t)^2 - (1-t)^2}{(1+t)^2 + (1-t)^2} = \frac{2t}{1+t^2}.$$

By the transfer formula,

$$t_n(x) = x\left(\frac{1+t^2}{2}\right)^n x^{n-1}$$

$$= \frac{1}{2^n} x \sum_{k=0}^{n} \binom{n}{k} t^{2k} x^{n-1}$$

$$= \frac{1}{2^n} \sum_{k \geq 0} \binom{n}{k} (n-1)_{2k} x^{n-2k},$$

and so

$$M_n\left(\frac{x}{2}\right) = \frac{1}{2^n} \sum_{k \geq 0} \binom{n}{k} (n-1)_{2k} M_{n-2k}(x).$$

2. THE LAGRANGE INVERSION FORMULA

The Lagrange inversion formula is a formula for the compositional inverse of a delta series. Actually, the most common version of the Lagrange formula is more general and there are variants. A concise treatment from the classical point of view, with references, can be found in Comtet [1].

Let $g(t)$ be any formal power series and let $f(t)$ be a delta series.

The Lagrange Inversion Formula   The coefficient of t^n in $g(\bar{f}(t))$, multiplied by n, equals the coefficient of t^{n-1} in $g'(t)(f(t)/t)^{-n}$. In the symbols of the umbral calculus,

$$\langle g(\bar{f}(t))\,|\,x^n\rangle = \left\langle g'(t)\left(\frac{f(t)}{t}\right)^{-n}\,\bigg|\,x^{n-1}\right\rangle.$$

Hermite's Version of the Lagrange Inversion Formula   The coefficient of t^n in $tg(\bar{f}(t))/\bar{f}(t)f'(\bar{f}(t))$ equals the coefficient of t^n in $g(t)(f(t)/t)^{-n}$. In symbols,

$$\left\langle \frac{tg(\bar{f}(t))}{\bar{f}(t)f'(\bar{f}(t))}\,\bigg|\,x^n\right\rangle = \left\langle g(t)\left(\frac{f(t)}{t}\right)^{-n}\,\bigg|\,x^n\right\rangle.$$

When $g(t) = t$, the Lagrange formula reduces to

$$\langle \bar{f}(t)\,|\,x^n\rangle = \left\langle \left(\frac{f(t)}{t}\right)^{-n}\,\bigg|\,x^{n-1}\right\rangle,$$

a formula for the compositional inverse of $f(t)$.

Notice that the key feature of these formulas, indeed of any variant of the Lagrange formula, is the presence of $\bar{f}(t)$ on one side and its absence on the other.

There have been many different proofs of these formulas, many of which assume that the power series involved are convergent and represent functions analytic in a disk (for an example see Whittaker and Watson [1]). Cauchy's theorem is frequently invoked. But, as Henrici [1] points out, the essential character of the Lagrange formula is purely formal.

Using umbral methods, the proof of the Lagrange inversion formula and its variants reduces to a simple computation. From the second form of the transfer formula we have

$$\begin{aligned}
\langle g(\bar{f}(t))\,|\,x^n\rangle &= \langle \lambda_f^* g(t)\,|\,x^n\rangle \\
&= \langle g(t)\,|\,\lambda_f x^n\rangle \\
&= \langle g(t)\,|\,x(f(t)/t)^{-n}x^{n-1}\rangle \\
&= \langle g'(t)\,|\,(f(t)/t)^{-n}x^{n-1}\rangle \\
&= \langle g'(t)(f(t)/t)^{-n}\,|\,x^{n-1}\rangle.
\end{aligned}$$

It is hard to imagine a simpler proof. Moreover, from this proof we see how to obtain variants of the formula. Using the first form of the transfer formula, we get

$$\langle g(\bar{f}(t)) \mid x^n \rangle = \langle g(t) \mid \lambda_f x^n \rangle$$
$$= \langle g(t) \mid f'(t)(f(t)/t)^{-n} x^n \rangle$$
$$= \langle g(t) f'(t)(f(t)/t)^{-n-1} \mid x^n \rangle.$$

Thus we have another version of Lagrange's formula.

A Third Variant of the Lagrange Inversion Formula   The coefficient of t^n in $g(\bar{f}(t))$ equals the coefficient of t^n in $g(t) f'(t)(f(t)/t)^{-n-1}$. In symbols,

$$\langle g(\bar{f}(t)) \mid x^n \rangle = \left\langle g(t) f'(t) \left(\frac{f(t)}{t} \right)^{-n-1} \middle| x^n \right\rangle.$$

If we replace $g(t)$ by $g(t)f(t)/tf'(t)$ we obtain Hermite's variant.

3. CROSS SEQUENCES AND STEFFENSEN SEQUENCES

The reader may have observed that many of the examples of the previous chapter share a common property. Namely, they have the form

$$s_n^{(\lambda)}(x) = h(t)^\lambda s_n(x),$$

where $s_n(x)$ is a Sheffer sequence, $h(t)$ is invertible, and λ ranges over the real numbers. The sequence $s_n^{(\lambda)}(x)$ is known as the *Steffensen sequence* for $h(t)$ and $s_n(x)$ of index λ. In case $s_n(x)$ is an associated sequence, then $s_n^{(\lambda)}(x)$ is known as the *cross sequence* for $h(t)$ and $s_n(x)$ of index λ. If $s_n(x) = x^n$, then $s_n^{(\lambda)}(x) = h(t)^\lambda x^n$ is the *Appell cross sequence* for $h(t)$ of index λ.

For example, the Hermite, Bernoulli (of the first kind), and Euler polynomials form Appell cross sequences. The Poisson–Charlier and actuarial polynomials form cross sequences, and the Laguerre polynomials form a Steffensen sequence.

Of course, any Sheffer sequence $s_n(x)$ generates, in a natural way, a cross sequence. For if $s_n(x)$ is Sheffer for $(g(t), f(t))$ and $p_n(x)$ is associated to $f(t)$, then

$$s_n(x) = g(t)^{-1} p_n(x),$$

and so

$$s_n^{(\lambda)}(x) = g(t)^\lambda p_n(x)$$

is a cross sequence for which

$$s_n^{(-1)}(x) = s_n(x).$$

The theory of Steffensen sequences per se has not been well developed (but see Rota *et al.* [1], and J. W. Brown [1]). We shall present here only a few results.

Theorem 5.3.1 A sequence $p_n^{(\lambda)}(x)$ is a cross sequence if and only if

$$p_n^{(\lambda+\mu)}(x + y) = \sum_{k=0}^{n} \binom{n}{k} p_k^{(\lambda)}(y) p_{n-k}^{(\mu)}(x) \tag{5.3.1}$$

for all $n \geq 0$, all y in C, and all real numbers λ and μ.

Proof Let $p_n^{(\lambda)}(x) = h(t)^\lambda p_n(x)$ be a cross sequence. If $p_n(x)$ is associated to $f(t)$, then $f(t)p_n^{(\lambda)}(x) = np_{n-1}^{(\lambda)}(x)$, and so for each λ the sequence $p_n^{(\lambda)}(x)$ is Sheffer and

$$p_n^{(\lambda)}(x + y) = \sum_{k=0}^{n} \binom{n}{k} p_k^{(\lambda)}(y) p_{n-k}(x).$$

Applying $h(t)^\mu$ to this equation gives (5.3.1). Conversely, if (5.3.1) holds, then taking $\mu = 0$ shows that $p_n^{(\lambda)}(x)$ is Sheffer, say for $(h_\lambda(t), f(t))$. Thus $p_n^{(\lambda)}(x) = h_\lambda(t)^{-1} p_n(x)$. Now

$$\langle h_\lambda(t)^{-1} h_\mu(t)^{-1} \mid p_n(x) \rangle = \sum_{k=0}^{n} \binom{n}{k} \langle h_\lambda(t)^{-1} \mid p_k(x) \rangle \langle h_\mu(t)^{-1} \mid p_{n-k}(x) \rangle$$

$$= \sum_{k=0}^{n} \binom{n}{k} p_k^{(\lambda)}(0) p_{n-k}^{(\mu)}(0)$$

$$= p_n^{(\lambda+\mu)}(0)$$

$$= \langle h_{\lambda+\mu}(t)^{-1} \mid p_n(x) \rangle,$$

and so $h_\lambda(t)h_\mu(t) = h_{\lambda+\mu}(t)$, which implies that $h_\lambda(t) = h(t)^\lambda$. Thus

$$p_n^{(\lambda)}(x) = h(t)^\lambda p_n(x)$$

is a cross sequence.

Theorem 5.3.2 A sequence $s_n^{(\lambda)}(x)$ is a Steffensen sequence if and only if

$$s_n^{(\lambda+\mu)}(x + y) = \sum_{k=0}^{n} \binom{n}{k} s_k^{(\lambda)}(x) p_{n-k}^{(\mu)}(y) \tag{5.3.2}$$

for some cross sequence $p_n^{(\mu)}(x)$.

Proof If $s_n^{(\lambda)}(x)$ is a Steffensen sequence, then $s_n^{(\lambda)}(x) = g(t)^{-1}p_n^{(\lambda)}(x)$, where $p_n^{(\lambda)}(x)$ is a cross sequence. Applying $g(t)^{-1}$ to (5.3.1) with x and y interchanged, we get (5.3.2). For the converse, taking $\mu = 0$ in (5.3.2), we deduce the existence, for each λ, of an operator $g_\lambda(t)$ for which $g_\lambda(t)s_n^{(\lambda)}(x) = p_n^{(\lambda)}(x)$. Then

$$g_{\lambda+\mu}(t)^{-1}p_n^{(\lambda+\mu)}(x+y) = s_n^{(\lambda+\mu)}(x+y)$$

$$= \sum_{k=0}^{n} \binom{n}{k} s_k^{(\lambda)}(x)p_{n-k}^{(\mu)}(y)$$

$$= g_\lambda(t)^{-1} \sum_{k=0}^{n} \binom{n}{k} p_k^{(\lambda)}(x)p_{n-k}^{(\mu)}(y)$$

$$= g_\lambda(t)^{-1}p_n^{(\lambda+\mu)}(x+y),$$

and so $g_{\lambda+\mu}(t) = g_\lambda(t)$, which implies that $g_\lambda(t) = g_0(t)$. Thus

$$s_n^{(\lambda)}(x) = g_0(t)^{-1}p_n^{(\lambda)}(x)$$

is a Steffensen sequence.

The umbral composition of Appell cross sequences is very simple.

Theorem 5.3.3 If $p_n^{(\lambda)}(x)$ is an Appell cross sequence, then

$$p_n^{(\lambda)}(\mathbf{p}^{(\mu)}(x)) = p_n^{(\lambda+\mu)}(x).$$

Proof This follows from the fact that the umbral operator for $p_n^{(\lambda)}(x)$ is $h(t)^\lambda$.

We conclude this section with some more exotic results.

Theorem 5.3.4 If $s_n^{(\lambda)}(x) = h(t)^\lambda s_n(x)$ is a Steffensen sequence, where $s_n(x)$ is Sheffer for $(g(t), f(t))$, then the sequence

$$s_n^{(an)}(x)$$

is Sheffer for

$$\left(\frac{(h(t)^{-a}f(t))'}{f'(t)} h(t)^a g(t), h(t)^{-a}f(t) \right).$$

Proof First we observe that

$$h(t)^{-a}f(t)s_n^{(an)}(x) = h(t)^{-a}f(t)h(t)^{an}s_n(x)$$

$$= nh(t)^{a(n-1)}s_{n-1}(x)$$

$$= ns_{n-1}^{(an-a)}(x).$$

Also, if $p_n(x) = g(t)s_n(x)$ is associated to $f(t)$, then

$$\frac{(h(t)^{-a}f(t))'}{f'(t)} h(t)^a g(t) s_n^{(an)}(x) = \frac{(h(t)^{-a}f(t))'}{f'(t)} h(t)^{a(n+1)} p_n(x)$$

$$= \frac{(h(t)^{-a}f(t))'}{f'(t)} h(t)^{a(n+1)} f'(t) \left(\frac{f(t)}{t}\right)^{-n-1} x^n$$

$$= (h(t)^{-a}f(t))' \left(\frac{h(t)^{-a}f(t)}{t}\right)^{-n-1} x^n,$$

which, by the transfer formula, is the associated sequence for $h(t)^{-a}f(t)$. This concludes the proof.

Theorem 5.3.5 If $s_n^{(\lambda)}(x) = h(t)^\lambda [f'(t)]^{-1} p_n(x)$ is a Steffensen sequence, where $p_n(x)$ is associated to $f(t)$, then the sequence

$$x s_{n-1}^{(an)}(x)$$

for $n \geq 1$, is associated to $h(t)^{-a}f(t)$.

Proof By the transfer formula and the recurrence formula we see that the associated sequence for $h(t)^{-a}f(t)$ is

$$x\left(\frac{h(t)^{-a}f(t)}{t}\right)^{-n} x^{n-1} = xh(t)^{an}\left(\frac{f(t)}{t}\right)^{-n} x^{n-1}$$

$$= xh(t)^{an} x^{-1} x \left(\frac{f(t)}{t}\right)^{-n} x^{n-1}$$

$$= xh(t)^{an} x^{-1} p_n(x)$$

$$= xh(t)^{an} x^{-1} x [f'(t)]^{-1} p_{n-1}(x)$$

$$= xh(t)^{an} [f'(t)]^{-1} p_{n-1}(x)$$

$$= x s_{n-1}^{(an)}(x).$$

Appell cross sequences fit the requirements of this theorem with $f(t) = t$.

Corollary 5.3.6 If $s_n^{(\lambda)}(x) = h(t)^\lambda x^n$ is an Appell cross sequence, then

$$x s_{n-1}^{(an)}(x)$$

is associated to $h(t)^{-a}t$.

Corollary 5.3.7 If $s_n^{(\lambda)}(x) = h(t)^\lambda [f'(t)]^{-1} p_n(x)$ is a Steffensen sequence, where $p_n(x)$ is associated to $f(t)$, then

$$(x+y)s_{n-1}^{(an)}(x+y) = \sum_{k=0}^{n} \binom{n}{k} xy s_{k-1}^{(ak)}(x) s_{n-k-1}^{(an-ak)}(y).$$

4. OPERATIONAL FORMULAS

In this section we present a simple umbral technique for deriving operational formulas. A well-known example of an operational formula is

$$(xt)^n = \sum_{k=0}^{n} S(n,k)x^k t^k, \tag{5.4.1}$$

where t is the derivative operator and $S(n,k)$ are the Stirling numbers of the second kind. Many formulas of this type are obtained by catch-as-catch-can methods and then proved by induction. The umbral approach gives a systematic method for generating such formulas.

Let $s_n(x)$ be Sheffer for $(g(t), f(t))$. Suppose that Q_k is a sequence of linear operators on P for which

$$Q_k s_n(x) = r_k(n)s_{n+\mu}(x),$$

where $r_n(x)$ is Sheffer for $(h(t), l(t))$ and μ is an integer. Suppose further that T is an operator on P for which

$$Ts_n(x) = u(n)s_{n+\lambda}(x),$$

where $u(x)$ is a polynomial and λ is an integer. We wish to expand T in terms of the operators Q_k.

By the polynomial expansion theorem,

$$u(x) = \sum_{k \geq 0} \frac{1}{k!} \langle h(t)l(t)^k \,|\, u(x) \rangle r_k(x).$$

If $\theta: s_n(x) \to s_{n+1}(x)$ is the Sheffer shift for $s_n(x)$ and if $\mu \geq \min\{0, \lambda\}$, then

$$Ts_n(x) = \sum_{k \geq 0} \frac{1}{k!} \langle h(t)l(t)^k \,|\, u(x) \rangle r_k(n)s_{n+\lambda}(x)$$

$$= \theta^{\lambda - \mu} \sum_{k \geq 0} \frac{1}{k!} \langle h(t)l(t)^k \,|\, u(x) \rangle r_k(n)s_{n+\mu}(x)$$

$$= \theta^{\lambda - \mu} \sum_{k \geq 0} \frac{1}{k!} \langle h(t)l(t)^k \,|\, u(x) \rangle Q_k s_n(x),$$

and so the desired expansion is

$$T = \theta^{\lambda - \mu} \sum_{k \geq 0} \frac{1}{k!} \langle h(t)l(t)^k \,|\, u(x) \rangle Q_k. \tag{5.4.2}$$

We have not taken the most general approach to deriving a formula of the type (5.4.2). In Roman [9] we consider more general operators Q_k, where $r_n(x)$

may not be Sheffer. However, to avoid details that tend to obscure the main idea, we shall confine ourselves here even to the case $\lambda = \mu = 0$.

Theorem 5.4.1   Let $s_n(x)$ be Sheffer for $(g(t), f(t))$ and let $r_n(x)$ be Sheffer for $(h(t), l(t))$. Then if Q_k and T are operators defined by

$$Q_k s_n(x) = r_k(n)s_n(x), \qquad Ts_n(x) = u(n)s_n(x),$$

where $u(x)$ is a polynomial, we have

$$T = \sum_{k \geq 0} \frac{1}{k!} \langle h(t)l(t)^k \mid u(x) \rangle Q_k.$$

Let us consider a simple example. If we take

$$Q_k = \theta^k f(t)^k,$$

then

$$Q_k s_n(x) = \theta^k f(t)^k s_n(x) = (n)_k s_n(x),$$

where, of course, if $k > n$, then $(n)_k = 0$. Thus

$$r_n(x) = (x)_n, \qquad h(t) = 1, \qquad l(t) = e^t - 1,$$

and so we have

$$T = \sum_{k \geq 0} \frac{1}{k!} \langle (e^t - 1)^k \mid u(x) \rangle \theta^k f(t)^k.$$

Taking

$$T = [\theta f(t)]^m,$$

we have

$$Ts_n(x) = [\theta f(t)]^m s_n(x) = n^m s_n(x),$$

and so

$$u(x) = x^m.$$

Thus we obtain the operational formula

$$[\theta f(t)]^m = \sum_{k=0}^{m} \frac{1}{k!} \langle (e^t - 1)^k \mid x^m \rangle \theta^k f(t)^k$$

or, since $(1/k!)\langle (e^t - 1)^k \mid x^m \rangle = S(m, k)$ are the Stirling numbers of the second kind,

$$[\theta f(t)]^m = \sum_{k=0}^{m} S(m, k)\theta^k f(t)^k. \qquad (5.4.3)$$

Here $\theta: s_n(x) \to s_{n+1}(x)$ is the Sheffer shift for $s_n(x)$, which is Sheffer for a pair whose delta series is $f(t)$.

Let us give some instances of (5.4.3).

Example 1 Let $s_n(x) = (x - a)^n$, which is Sheffer for (e^{at}, t). By Theorem 3.7.1,

$$\theta = \left(x - \frac{g'(t)}{g(t)} \right) \frac{1}{f'(t)} = x - a,$$

and (5.4.3) becomes

$$[(x - a)t]^m = \sum_{k=0}^{m} S(m, k)(x - a)^k t^k,$$

which, for $a = 0$, is (5.4.1).

Example 2 Let $s_n(x) = (x/a)(x/(a + 1)) \cdots (x/(a + n - 1))$, which is associated to the backward difference $f(t) = 1 - e^{-at}$. Then

$$\theta = x[f'(t)]^{-1} = a^{-1} x e^{at},$$

and so we get

$$[a^{-1} x (e^{at} - 1)]^m = \sum_{k=0}^{m} S(m, k) a^{-k} (x e^{at})^k (1 - e^{-at})^k.$$

Using the notation

$$\Delta_a = e^{at} - 1,$$

$$\nabla_a = 1 - e^{-at},$$

$$E^a = e^{at}$$

for the forward difference, backward difference, and translation operators, we have

$$(x \Delta_a)^m = \sum_{k=0}^{m} S(m, k) a^{m-k} (x E^a)^k \nabla_a^k.$$

Example 3 Let $s_n(x) = H_n^{(v)}(x)$ be the Hermite polynomials. Then

$$\theta = x - \frac{g'(t)}{g(t)} = x + vt,$$

and so we get

$$[(x + vt)t]^m = \sum_{k=0}^{m} S(m, k)(x + vt)^k t^k.$$

Other examples of Theorem 5.4.1 are just as easily obtained.

5. INVERSE RELATIONS

An inverse relation is a pair of equations of the form

$$y_n = \sum_{k=0}^{n} a_{n,k} x_k,$$

$$x_n = \sum_{k=0}^{n} b_{n,k} y_k,$$

(5.5.1)

where x_n and y_n are variables. One of the simplest examples of an inverse relation is

$$y_n = \sum_{k=0}^{n} \binom{n}{k} x_k,$$

$$x_n = \sum_{k=0}^{n} (-1)^{n-k} \binom{n}{k} y_k.$$

There is a vast literature on inverse relations, and Riordan [2] devotes no less than 85 pages to the subject.

The relations (5.5.1) are equivalent to a single orthogonality relation, obtained by substitution,

$$\delta_{n,j} = \sum_{k=j}^{n} a_{n,k} b_{k,j}.$$

(5.5.2)

From our point of view, we observe that if $s_n(x)$ and $r_n(x)$ are *sequences* of polynomials for which

$$r_n(x) = \sum_{k=0}^{n} a_{n,k} s_k(x)$$

and

$$s_n(x) = \sum_{k=0}^{n} b_{n,k} r_k(x),$$

then $a_{n,k}$ and $b_{n,k}$ satisfy (5.5.2), and so also (5.5.1).

Now suppose that $s_n(x)$ is Sheffer for $(g(t), f(t))$ and $r_n(x)$ is Sheffer for $(h(t), l(t))$. Then by the polynomial expansion theorem

$$s_n(x) = \sum_{k=0}^{n} \frac{\langle h(t)l(t)^k \mid s_n(x) \rangle}{k!} r_k(x)$$

and

$$r_n(x) = \sum_{k=0}^{n} \frac{\langle g(t)f(t)^k \mid r_n(x) \rangle}{k!} s_k(x).$$

This gives the inverse pair

$$y_n = \sum_{k=0}^{n} \frac{\langle h(t)l(t)^k \,|\, s_n(x) \rangle}{k!} x_k,$$

$$x_n = \sum_{k=0}^{n} \frac{\langle g(t)f(t)^k \,|\, r_n(x) \rangle}{k!} y_k.$$

Actually, this pair can be simplified somewhat. For if $p_n(x)$ is associated to $f(t)$ and $q_n(x)$ is associated to $l(t)$, then

$$y_n = \sum_{k=0}^{n} \frac{1}{k!} \langle g(t)^{-1} h(t)l(t)^k \,|\, p_n(x) \rangle x_k,$$

$$x_n = \sum_{k=0}^{n} \frac{1}{k!} \langle h(t)^{-1} g(t)f(t)^k \,|\, q_n(x) \rangle y_k.$$

Since $g(t)$ and $h(t)$ are arbitrary (invertible) series, we have proved the following result.

Theorem 5.5.1 Let $p_n(x)$ be associated to $f(t)$ and let $q_n(x)$ be associated to $l(t)$. Then for any invertible series $h(t)$ we have the inverse pair

$$y_n = \sum_{k=0}^{n} \frac{\langle h(t)l(t)^k \,|\, p_n(x) \rangle}{k!} x_k,$$

$$x_n = \sum_{k=0}^{n} \frac{\langle h(t)^{-1} f(t)^k \,|\, q_n(x) \rangle}{k!} y_k.$$

In case $f(t) = l(t)$, this theorem can be simplified.

Corollary 5.5.2 Let $p_n(x)$ be an associated sequence. Then for any invertible $h(t)$ we have the inverse pair

$$y_n = \sum_{k=0}^{n} \binom{n}{k} \langle h(t) \,|\, p_{n-k}(x) \rangle x_k,$$

$$x_n = \sum_{k=0}^{n} \binom{n}{k} \langle h(t)^{-1} \,|\, p_{n-k}(x) \rangle y_k.$$

Corollary 5.5.3 Let $p_n(x)$ be an associated sequence. Then for any constant y we have

$$y_n = \sum_{k=0}^{n} \binom{n}{k} p_{n-k}(y) x_k,$$

$$x_n = \sum_{k=0}^{n} \binom{n}{k} p_{n-k}(-y) y_k.$$

Let us give a few examples.

Example 1   If $p_n(x) = x^n$, we get

$$y_n = \sum_{k=0}^{n} \binom{n}{k} y^{n-k} x_k,$$

$$x_n = \sum_{k=0}^{n} \binom{n}{k} (-y)^{n-k} y_k.$$

Example 2   If $p_n(x) = (x)_n$, then we have

$$y_n = \sum_{k=0}^{n} \binom{n}{k} (y)_{n-k} x_k,$$

$$x_n = \sum_{k=0}^{n} \binom{n}{k} (-y)_{n-k} y_k.$$

Example 3   If $p_n(x) = G_n(x;a;b)$ are the Gould polynomials, we get

$$y_n = \sum_{k=0}^{n} \frac{y}{y - a(n-k)} \left(\frac{y - a(n-k)}{b} \right)_{n-k} x_k,$$

$$x_n = \sum_{k=0}^{n} \frac{y}{y + a(n-k)} \left(\frac{-y - a(n-k)}{b} \right)_{n-k} y_k.$$

Example 4   The Fibonacci numbers F_n can be defined by the generating function

$$\sum_{k=0}^{\infty} F_k t^k = \frac{1}{1 - t - t^2}.$$

If $p_n(x)$ is the associated sequence for which

$$\sum_{k=0}^{\infty} \frac{p_k(x)}{k!} t^k = (1 - t - t^2)^{-x},$$

then, for $x = 1$, we get

$$\frac{p_k(1)}{k!} = F_k$$

and, for $x = -1$, we get

$$p_0(-1) = 1,$$
$$p_1(-1) = -1,$$
$$p_2(-1) = -2,$$
$$p_k(-1) = 0, \qquad k > 2.$$

Thus, taking $y = -1$ in Corollary 5.5.3, we get the inverse pair

$$y_n = x_n - nx_{n-1} - n(n-1)x_{n-2},$$

$$x_n = \sum_{k=0}^{n} \frac{n!}{k!} F_{n-k} y_k.$$

Replacing x_n by $x_n/n!$ and y_n by $y_n/n!$ gives

$$y_n = x_n - x_{n-1} - x_{n-2},$$

$$x_n = \sum_{k=0}^{n} F_{n-k} y_k.$$

Next, let us take $h(t) = u(t)l(t)^\lambda$ in Corollary 5.5.2, giving

$$y_n = \sum_{k=0}^{n} \binom{n}{k} \langle u(t) \mid s_{n-k}^{(\lambda)}(x) \rangle x_k,$$

$$x_n = \sum_{k=0}^{n} \binom{n}{k} \langle u(t)^{-1} \mid s_{n-k}^{(-\lambda)}(x) \rangle y_k,$$

(5.5.3)

where $s_n^{(\lambda)}(x) = l(t)^\lambda p_n(x)$ is, in the language of Section 3, a cross sequence. Taking $u(t) = e^{yt}$, we get

$$y_n = \sum_{k=0}^{n} \binom{n}{k} s_{n-k}^{(\lambda)}(y) x_k,$$

$$x_n = \sum_{k=0}^{n} \binom{n}{k} s_{n-k}^{(-\lambda)}(-y) y_k.$$

We continue the examples.

Example 5   Let $s_n^{(\lambda)}(x) = H_n^{(\lambda)}(x)$ be the Hermite polynomials. Then

$$y_n = \sum_{k=0}^{n} \binom{n}{k} H_{n-k}^{(\lambda)}(y) x_k,$$

$$x_n = \sum_{k=0}^{n} \binom{n}{k} H_{n-k}^{(-\lambda)}(-y) y_k.$$

Since

$$H_{n-k}^{(\lambda)}(0) = \begin{cases} \left(\dfrac{-\lambda}{2}\right)^{(n-k)/2} \dfrac{(n-k)!}{\left(\frac{1}{2}(n-k)\right)!}, & n-k \quad \text{even,} \\ 0, & n-k \quad \text{odd,} \end{cases}$$

we get the inverse pair

$$y_n = \sum_{\substack{k=0 \\ n-k \text{ even}}}^{n} \binom{n}{k} \left(-\frac{\lambda}{2}\right)^{(n-k)/2} \frac{(n-k)!}{(\frac{1}{2}(n-k))!} x_k,$$

$$x_n = \sum_{\substack{k=0 \\ n-k \text{ even}}}^{n} \binom{n}{k} \left(\frac{\lambda}{2}\right)^{(n-k)/2} \frac{(n-k)!}{(\frac{1}{2}(n-k))!} y_k.$$

Example 6   If $s_n^{(\lambda)}(x) = B_n^{(\lambda)}(x)$ are the Bernoulli polynomials, we have

$$y_n = \sum_{k=0}^{n} \binom{n}{k} B_{n-k}^{(\lambda)}(y) x_k,$$

$$x_n = \sum_{k=0}^{n} \binom{n}{k} B_{n-k}^{(-\lambda)}(-y) y_k.$$

For $\lambda = 1$ and $y = 0$, since

$$B_{n-k}^{(-1)}(0) = \left\langle \frac{e^t - 1}{t} \,\middle|\, x^{n-k} \right\rangle = \frac{1}{n-k+1},$$

we have

$$y_n = \sum_{k=0}^{n} \binom{n}{k} B_{n-k}(0) x_k,$$

$$x_n = \sum_{k=0}^{n} \binom{n}{k} \frac{1}{n-k+1} y_k.$$

Example 7   The Laguerre polynomials satisfy

$$L_n^{(\lambda)}(x) = (1-t)^{\lambda+1} L_n(x),$$

and so if we set $u(t) = e^{yt}(1-t)$ and $s_n^{(\lambda)}(x) = (1-t)^{\lambda} L_n(x)$ in (5.5.3), we get

$$y_n = \sum_{k=0}^{n} \binom{n}{k} L_{n-k}^{(\lambda)}(y) x_k,$$

$$x_n = \sum_{k=0}^{n} \binom{n}{k} L_{n-k}^{(-\lambda-2)}(-y) y_k.$$

Example 8   Recalling that the Laguerre polynomials are self-inverse, that is,

$$L_n^{(\lambda)}\big(\mathbf{L}^{(\lambda)}(x)\big) = x^n,$$

we get the orthogonality relation

$$x^n = \sum_{k=0}^{n} \frac{n!}{k!} \binom{\lambda + n}{\lambda + k}(-1)^k L_k^{(\lambda)}(x)$$

$$= \sum_{k=0}^{n} \frac{n!}{k!} \binom{\lambda + n}{\lambda + k}(-1)^k \sum_{j=0}^{k} \frac{k!}{j!} \binom{\lambda + k}{\lambda + j}(-1)^j x^j$$

$$= \sum_{j=0}^{n} \sum_{k=j}^{n} \frac{n!}{j!} \binom{\lambda + n}{\lambda + k}\binom{\lambda + k}{\lambda + j}(-1)^{k+j} x^j,$$

which gives the inverse pair

$$y_n = \sum_{k=0}^{n} \frac{n!}{k!} \binom{\lambda + n}{\lambda + k}(-1)^k x_k,$$

$$x_n = \sum_{k=0}^{n} \frac{n!}{k!} \binom{\lambda + n}{\lambda + k}(-1)^k y_k.$$

Let us consider a somewhat more delicate corollary of Theorem 5.5.1. Namely, we take

$$f(t) = e^{at} u(t) \quad \text{and} \quad l(t) = e^{bt} u(t),$$

where $u(t)$ is a delta series and $a \neq b$. Then Theorem 5.5.1 gives the inverse pair whose coefficients are

$$\frac{1}{k!} \langle h(t) e^{bkt} u(t)^k \mid p_n(x) \rangle,$$

$$\frac{1}{k!} \langle h(t)^{-1} e^{akt} u(t)^k \mid q_n(x) \rangle.$$

This is equivalent to the pair of coefficients

$$\frac{1}{k!} \langle h(t) e^{(b-a)kt} \mid [e^{at} u(t)]^k p_n(x) \rangle,$$

$$\frac{1}{k!} \langle h(t)^{-1} e^{(a-b)kt} \mid [e^{bt} u(t)]^k q_n(x) \rangle,$$

which in turn is equivalent to

$$\binom{n}{k} \langle h(t) e^{(b-a)kt} \mid p_{n-k}(x) \rangle,$$

$$\binom{n}{k} \langle h(t)^{-1} e^{(a-b)kt} \mid q_{n-k}(x) \rangle$$

or

$$\binom{n}{k}\langle h(t) \mid p_{n-k}(x + (b-a)k)\rangle,$$

$$\binom{n}{k}\langle h(t)^{-1} \mid q_{n-k}(x + (a-b)k)\rangle. \tag{5.5.4}$$

Now if $u_n(x)$ is associated to $u(t)$, then, according to Corollary 3.8.2,

$$p_n(x) = x\left(\frac{u(t)}{e^{at}u(t)}\right)^n x^{-1}u_n(x)$$

$$= xe^{-ant}x^{-1}u_n(x)$$

$$= \frac{x}{x - an}u_n(x - an),$$

and, similarly,

$$q_n(x) = \frac{x}{x - bn}u_n(x - bn).$$

Putting this into (5.5.4) gives the following result.

Corollary 5.5.4 Let $u_n(x)$ be an associated sequence. Then for any invertible $h(t)$ we have the inverse pair

$$y_n = \sum_{k=0}^{n}\binom{n}{k}\left\langle h(t) \left| \frac{x + (b-a)k}{x - an + bk}u_{n-k}(x - an + bk)\right.\right\rangle x_k,$$

$$x_n = \sum_{k=0}^{n}\binom{n}{k}\left\langle h(t)^{-1} \left| \frac{x + (a-b)k}{x - bn + ak}u_{n-k}(x - bn + ak)\right.\right\rangle y_k.$$

Corollary 5.5.5 Let $u_n(x)$ be an associated sequence. Then for any constant y we have

$$y_n = \sum_{k=0}^{n}\binom{n}{k}\frac{y + (b-a)k}{y - an + bk}u_{n-k}(y - an + bk)x_k,$$

$$x_n = \sum_{k=0}^{n}\binom{n}{k}\frac{-y + (a-b)k}{-y - bn + ak}u_{n-k}(-y - bn + ak)y_k.$$

Let us consider some examples of Corollary 5.5.5.

Example 9 If we take $u_n(x) = (x)_n$ and $a = 0$, we get

$$y_n = \sum_{k=0}^{n}\binom{n}{k}(y + bk)_{n-k}x_k,$$

$$x_n = \sum_{k=0}^{n}\binom{n}{k}\frac{y + bk}{y + bn}(-y - bn)_{n-k}y_k,$$

and when $b = 0$,

$$y_n = \sum_{k=0}^{n} \frac{n!}{k!}\binom{y}{n-k}x_k,$$

$$x_n = \sum_{k=0}^{n} \frac{n!}{k!}\binom{-y}{n-k}y_k.$$

Example 10   If we take $u_n(x) = (x)_n$ and $b = -1$, we get

$$y_n = \sum_{k=0}^{n} \frac{n!}{k!} \frac{y - (a+1)k}{y - an - k}\binom{y - an - k}{n-k}x_k,$$

$$x_n = \sum_{k=0}^{n} \frac{n!}{k!} \frac{-y + (a+1)k}{-y + n + ak}\binom{-y + n + ak}{n-k}y_k,$$

which is equivalent to a class of inverse relations given by Gould (Riordan [2, p. 50, Eq. (7)]).

Riordan [2, pp. 92–99] also gives four inverse relations associated with the name of Abel. If we take $u_n(x) = x^n$ in Corollary 5.5.5, we have

$$y_n = \sum_{k=0}^{n} \binom{n}{k}(y + (b-a)k)(y - an + bk)^{n-k-1}x_k,$$

$$x_n = \sum_{k=0}^{n} \binom{n}{k}(-y + (a-b)k)(-y - bn + ak)^{n-k-1}y_k.$$

Example 11   If $a = b = -1$, we get

$$y_n = \sum_{k=0}^{n} \binom{n}{k}y(y + n - k)^{n-k-1}x_k,$$

$$x_n = \sum_{k=0}^{n} \binom{n}{k}(-y)(-y + n - k)^{n-k-1}y_k,$$

which is Riordan's first Abel inverse relation.

Example 12   If $a = 0$ and $b = -1$, we have

$$y_n = \sum_{k=0}^{n} \binom{n}{k}(y - k)^{n-k}x_k,$$

$$x_n = \sum_{k=0}^{n} \binom{n}{k}(-y + k)(y + n)^{n-k-1}y_k,$$

which is Riordan's third Abel inverse relation.

Example 13 If $a = -1$ and $b = 1$, we get

$$y_n = \sum_{k=0}^{n} \binom{n}{k}(y + 2k)(y + n + k)^{n-k-1} x_k,$$

$$x_n = \sum_{k=0}^{n} \binom{n}{k}(-y - 2k)(-y - n - k)^{n-k-1} y_k,$$

which is Riordan's fourth Abel inverse relation.

To obtain Riordan's second Abel inverse relation, it is more instructive to invert one of the two relations rather than to prove that the given pair of relations are inverse to each other. In fact, we can generalize a bit and consider the problem of inverting the relation

$$y_n = \sum_{k=0}^{n} \binom{n}{k}(y + \alpha + \beta(n - k))^{n-k} x_k. \tag{5.5.5}$$

(When $\alpha = 0$ and $\beta = 1$, we get the first equation of Riordan's pair.) If we set

$$s_n(x) = (x + \alpha + \beta n)^n,$$

then, according to Theorem 3.8.3, $s_n(x)$ is Sheffer for

$$\left(e^{-\alpha t}(1 - \beta t), t e^{-\beta t}\right).$$

Thus

$$s_n(x) = e^{\alpha t}(1 - \beta t)^{-1} A_n(x; -\beta),$$

where $A_n(x; -\beta) = x(x + \beta n)^{n-1}$ are the Abel polynomials. Now (5.5.5) can be written as

$$y_n = \sum_{k=0}^{n} \binom{n}{k} \langle e^{yt} e^{\alpha t}(1 - \beta t)^{-1} \mid A_{n-k}(x; -\beta) \rangle x_k,$$

and by Corollary 5.5.2 its inverse is

$$x_n = \sum_{k=0}^{n} \binom{n}{k} \langle e^{-yt} e^{-\alpha t}(1 - \beta t) \mid A_{n-k}(x; -\beta) \rangle y_k.$$

It is straightforward to compute that

$$(1 - \beta t) A_m(x; -\beta) = (1 - \beta t)x(x + \beta m)^{m-1} = (x^2 - \beta^2 m)(x + \beta m)^{m-2},$$

and thus we get the inverse pair

$$y_n = \sum_{k=0}^{n} \binom{n}{k}(y + \alpha + \beta(n - k))^{n-k} x_k,$$

$$x_n = \sum_{k=0}^{n} \binom{n}{k}[(y + \alpha)^2 - \beta^2(n - k)][-y - \alpha + \beta(n - k)]^{n-k-2} y_k.$$

Setting $\alpha = 0$ and $\beta = 1$, we get Riordan's second Abel inverse relation

$$y_n = \sum_{k=0}^{n} \binom{n}{k}(y + n - k)^{n-k}x_k,$$

$$x_n = \sum_{k=0}^{n} \binom{n}{k}(y^2 - n + k)(-y + n - k)^{n-k-2}y_k.$$

The examples in this section are only a small sampling of the power of Theorem 5.5.1.

6. SHEFFER SEQUENCE SOLUTIONS TO RECURRENCE RELATIONS

The umbral calculus can be used to determine the Sheffer sequence solutions to certain recurrence relations. One of the most important nontrivial examples is that of the familiar three term recurrence

$$s_{n+1}(x) = (x - b_n)s_n(x) - d_n s_{n-1}(x) \qquad (5.6.1)$$

for $n \geq 0$, with $s_0(x) = 1$ and $s_{-1}(x) = 0$, which is known to characterize polynomial sequences that are orthogonal with respect to a moment functional. For details on orthogonality we refer the reader to Chihara [1].

The Sheffer sequence solutions to (5.6.1) were first determined by Meixner [1] in 1934 and then by Sheffer [1], using a different method, in 1939. The modern umbral method gives a very elementary solution to this and other recurrence relations.

The basic idea is that a Sheffer sequence satisfies a certain recurrence relation if and only if its associated pair of linear functionals satisfies a certain pair of formal differential equations.

Theorem 5.6.1 Let $h(t)$ be an invertible series and let $l(t)$ be any series. Then the recurrence relation

$$s_{n+1}(x) = [xh(t) + l(t)]s_n(x) + \sum_{j=0}^{m} n!\, c_{n,j} s_{n-j}(x) \qquad (5.6.2)$$

for $n \geq 0$, with $s_k(x) = 0$ for $k < 0$, has a solution $s_n(x)$, which is Sheffer for $(g(t), f(t))$, if and only if

(i) $o(g(t)) = 0, o(f(t)) = 1$,

(ii) $c_{k+j,j} = (1/k!)(kc_{j+1,j} - (k-1)c_{j,j})$ for $k \geq 1$,

(iii) $f'(t) = (1/h(t))[1 - \sum_{j=0}^{m}(c_{j+1,j} - c_{j,j})f(t)^{j+1}]$, and

(iv) $(g'(t)/g(t)) = -(1/h(t))[l(t) + \sum_{j=0}^{m} c_{j,j} f(t)^j]$.

Proof   Equation (5.6.2) holds for $s_n(x)$, Sheffer for $(g(t), f(t))$, if and only if it holds when $g(t)f(t)^k$ is applied to both sides, for all $k \geq 0$. Now

$$\langle g(t)f(t)^k \,|\, s_{n+1}(x)\rangle = (n+1)!\,\delta_{n+1,k}$$
$$= \langle kg(t)f(t)^{k-1}\,|\,s_n(x)\rangle,$$
$$\langle g(t)f(t)^k\,|\,[xh(t)+l(t)]s_n(x)\rangle = \langle [h(t)\,\partial_t + l(t)]g(t)f(t)^k\,|\,s_n(x)\rangle,$$

and

$$\langle g(t)f(t)^k\,|\,n!\,c_{n,j}s_{n-j}(x)\rangle = n!\,c_{n,j}k!\,\delta_{n,n-j}$$
$$= \langle k!\,c_{k+j,j}g(t)f(t)^{k+j}\,|\,s_n(x)\rangle,$$

which holds even for $n - j < 0$. Thus (5.6.2) holds for $s_n(x)$, Sheffer for $(g(t), f(t))$, if and only if

$$kg(t)f(t)^{k-1} = [h(t)\,\partial_t + l(t)]g(t)f(t)^k + \sum_{j=0}^{m} k!\,c_{k+j,j}g(t)f(t)^{k+j}$$

for all $k \geq 0$, along with $o(g(t)) = 0$ and $o(f(t)) = 1$. Taking the indicated derivative and canceling a factor $f(t)^{k-1}$ give

$$kg(t) = h(t)g'(t)f(t) + kh(t)g(t)f'(t) + l(t)g(t)f(t)$$
$$+ \sum_{j=0}^{m} k!\,c_{k+j,j}g(t)f(t)^{j+1} \tag{5.6.3}$$

for all $k \geq 0$. For $k = 0$ this is

$$0 = h(t)g'(t)f(t) + l(t)g(t)f(t) + \sum_{j=0}^{m} c_{j,j}g(t)f(t)^{j+1}. \tag{5.6.4}$$

By canceling a factor $f(t)$ and rearranging, we get (iv). Thus (5.6.2) holds if and only if (i), (5.6.4), and (5.6.3) hold.

Next we subtract (5.6.4) from (5.6.3), cancel a factor $g(t)$, and rearrange to get

$$h(t)f'(t) = 1 - \sum_{j=0}^{m} \frac{1}{k}(k!\,c_{k+j,j} - c_{j,j})f(t)^{j+1} \tag{5.6.5}$$

for all $k > 0$. Thus (5.6.2) is equivalent to (i), (5.6.4), and (5.6.5) for all $k > 0$. But (5.6.5) has a solution $f(t)$ for all $k > 0$ if and only if it has a solution when $k = 1$, which gives (iii), and

$$(1/k)(k!\,c_{k+j,j} - c_{j,j}) = c_{1+j,j} - c_{j,j}$$

for all $k > 0$, which gives (ii). Thus (5.6.2) is equivalent to (i)–(iv) and the proof is complete.

Condition (ii) is subject to separation of variables and can be solved in a variety of cases. Once $f(t)$ is determined, (iv) can be solved and, since $\langle g(t) | s_0(x) \rangle = 1$, we get

$$g(t) = \frac{1}{s_0(x)} \exp\left(-\int \left[l(t) + \sum_{j=0}^{m} c_{j,j} f(t)^j \right] \frac{1}{h(t)} dt \right). \qquad (5.6.6)$$

We may also obtain differential equations for $\bar{f}(t)$ and $g(\bar{f}(t))^{-1}$, for use in determining the generating function for the solution $s_n(x)$. Condition (iii) is equivalent to

$$\frac{[\bar{f}(t)]'}{h(\bar{f}(t))} = \left[1 - \sum_{j=0}^{m} (c_{j+1,j} - c_{j,j}) t^{j+1} \right]^{-1}, \qquad (5.6.7)$$

and from this and (5.6.6) we get

$$\frac{1}{g(\bar{f}(t))} = s_0(x) \exp\left(\int \frac{l(\bar{f}(t)) + \sum_{j=0}^{m} c_{j,j} t^j}{1 - \sum_{j=0}^{m}(c_{j+1,j} - c_{j,j}) t^{j+1}} dt \right). \qquad (5.6.8)$$

Corollary 5.6.2 The recurrence relation (5.6.2) has a solution $s_n(x)$, which is Sheffer for $(g(t), f(t))$, if and only if Eqs. (i), (ii), (5.6.7), and (5.6.8) hold.

Let us give two simple examples.

Example 1 The recurrence relation

$$s_{n+1}(x) = x s_n(x) - v s_n'(x),$$

with $s_0(x) = 1$, is well known to characterize the Hermite polynomials. Pretending ignorance of this fact and referring to Theorem 5.6.1, we have

$$h(t) = 1, \qquad l(t) = -vt, \qquad c_{n,k} = 0,$$

and so (iii) and (iv) give

$$f'(t) = 1 \qquad \text{and} \qquad g'(t)/g(t) = vt.$$

Hence

$$f(t) = t \qquad \text{and} \qquad g(t) = e^{vt^2/2},$$

which is indeed the pair for the Hermite polynomials.

Example 2 For the recurrence

$$s_{n+1}(x) = (2n + \alpha + 1 - x) s_n(x) - n(n + \alpha) s_{n-1}(x)$$

with $s_0(x) = 1$, we have

$$h(t) = -1, \qquad l(t) = \alpha + 1, \qquad m = 1,$$

$$c_{0,0} = 0, \qquad c_{n,0} = \frac{2}{(n-1)!} \quad (n > 0), \qquad c_{n,1} = -\frac{n+\alpha}{(n-1)!}.$$

Thus (iii) is

$$f'(t) = -\left(1 + f(t)\right)^2,$$

and so

$$f(t) = 1/(t - 1)$$

and (iv) is

$$\frac{g'(t)}{g(t)} = (\alpha + 1)(f(t) - 1) = \frac{\alpha + 1}{t - 1},$$

and so

$$g(t) = (1 - t)^{-\alpha - 1},$$

reminding us that the Laguerre polynomials satisfy this recurrence.

From these examples the reader should have no difficulty solving other recurrence relations (whose solutions are not known in advance).

Finally, let us consider (5.6.1), which is only a little less trivial than the preceding examples. In this case we have

$$h(t) = 1, \qquad l(t) = 0, \qquad m = 2,$$

$$c_{n, 0} = -\frac{b_n}{n!}, \qquad c_{n, 1} = -\frac{d_n}{n!}.$$

Equation (ii) gives, for $j = 0$ and then $j = 1$,

$$b_k = k(b_1 - b_0) + b_0$$

and

$$d_{k+1} = (k + 1)[k(d_2/2 - d_1) + d_1].$$

Equations (5.6.7) and (5.6.8) are

$$\bar{f}(t) = \int \frac{dt}{1 + (b_1 - b_0)t + (d_2/2 - d_1)t^2}$$

and

$$\frac{1}{g(\bar{f}(t))} = \exp \int \frac{-b_0 - d_1 t}{1 + (b_1 - b_0)t + (d_2/2 - d_1)t^2} \, dt.$$

These two integrals are routinely evaluated by elementary techniques and we omit the details. Suffice it to say that one obtains, as Sheffer did, four cases, depending on whether the denominator $1 + (b_1 - b_0)t + (d_2/2 - d_1)t^2$ is constant, linear, quadratic with distinct roots, or quadratic with confluent roots. Meixner considers five cases according to whether the denominator is constant, linear, or quadratic with negative discriminant, zero discriminant, or positive discriminant. Chihara [1, p. 164] gives the explicit solutions, which are in terms of the Hermite, Laguerre, Poisson–Charlier, and Meixner (of both kinds) polynomials (see also Erdelyi [1, Vol. 3, p. 273]).

The technique used in the proof of Theorem 5.6.1 works for types of recurrence relations other than the one given in that theorem. Thus if the reader has a particular recurrence relation he wishes to solve for Sheffer sequence solutions, this approach might prove useful.

7. BINOMIAL CONVOLUTION

We know that if $f(t)$ and $g(t)$ are delta series, then so is $f(g(t))$. Theorem 3.4.6 gives us the relationship between the associated sequences for $f(t), g(t)$ and $f(g(t))$.

Now if $f(t)$ and $g(t)$ are delta series satisfying $f'(0) + g'(0) \neq 0$, then $f(t) + g(t)$ is also a delta series. In this section we briefly consider the relationship between the associated sequences for $f(t), g(t)$, and $f(t) + g(t)$.

Theorem 5.7.1 Let $p_n(x)$ be associated to $f(t)$ and let $q_n(x)$ be associated to $g(t)$. If $f'(0) + g'(0) \neq 0$, then $\overline{f}(t) + \overline{g}(t)$ is a delta series whose associated sequence is

$$r_n(x) = \sum_{k=0}^{n} \binom{n}{k} p_k(x) q_{n-k}(x).$$

Proof   This follows from

$$e^{x(\overline{f}(t) + \overline{g}(t))} = e^{x\overline{f}(t)} e^{x\overline{g}(t)}$$

$$= \sum_{k=0}^{\infty} \frac{p_k(x)}{k!} t^k \sum_{k=0}^{\infty} \frac{q_k(x)}{k!} t^k$$

$$= \sum_{n=0}^{\infty} \left[\sum_{k=0}^{n} \binom{n}{k} p_k(x) q_{n-k}(x) \right] \frac{t^n}{n!}.$$

We call the sequence $r_n(x)$ in Theorem 5.7.1 the *binomial convolution of* $p_n(x)$ *and* $q_n(x)$, and write

$$p_n(x) \circ q_n(x) = \sum_{k=0}^{n} \binom{n}{k} p_k(x) q_{n-k}(x).$$

If we introduce the notation $\bar{p}_n(x)$ for the inverse, under umbral composition, of $p_n(x)$, then Theorem 5.7.1 has the following corollary.

Corollary 5.7.2 Let $p_n(x)$ be associated to $f(t)$ and let $q_n(x)$ be associated to $g(t)$. If $f'(0) + g'(0) \neq 0$, then the delta series $f(t) + g(t)$ has associated sequence

$$\overline{p_n(x) \circ q_n(x)}.$$

Theorem 5.7.1 can easily be extended to arbitrary Sheffer sequences, although the notation is a bit cumbersome.

Theorem 5.7.3 Let $s_n(x)$ be Sheffer for $(g(t), f(t))$ and let $r_n(x)$ be Sheffer for $(h(t), l(t))$. If $f'(0) + g'(0) \neq 0$, then the binomial convolution

$$s_n(x) \circ r_n(x) = \sum_{k=0}^{n} \binom{n}{k} s_k(x) \circ r_{n-k}(x)$$

is the Sheffer sequence for

$$(g(\overline{f}(\overline{f}(t) + \overline{g}(t)))h(\overline{l}(\overline{f}(t) + \overline{l}(t))), \overline{f}(t) + \overline{l}(t)).$$

In the Appell case Theorem 5.7.3 is a bit more manageable.

Corollary 5.7.4 Let $s_n(x)$ be Appell for $g(t)$ and let $r_n(x)$ be Appell for $h(t)$. Then the binomial convolution

$$s_n(x) \circ r_n(x) = \sum_{k=0}^{n} \binom{n}{k} s_k(x) r_{n-k}(x)$$

is Sheffer for

$$(g(t/2)h(t/2), t/2).$$

Thus

$$s_n(x) \circ r_n(x) = 2^n g(t/2)^{-1} h(t/2)^{-1} x^n.$$

As a simple example, the Bernoulli polynomials $B_n^{(a)}(x)$ are Appell for $(e^t - 1)^a / t^a$ and the Euler polynomials $E_n^{(a)}(x)$ are Appell for $(e^t + 1)^a / 2^a$. Thus

$$B_n^{(a)}(x) \circ E_n^{(a)}(x) = 2^n \left(\frac{t/2}{e^{t/2} - 1} \right)^a \left(\frac{2}{e^{t/2} + 1} \right)^a x^n$$

$$= 2^n \left(\frac{t}{e^t - 1} \right)^a x^n$$

$$= 2^n B_n^{(a)}(x).$$

That is,

$$\sum_{k=0}^{n} \binom{n}{k} B_k^{(a)}(x) E_{n-k}^{(a)}(x) = 2^n B_n^{(a)}(x).$$

CHAPTER 6

NONCLASSICAL UMBRAL CALCULI

1. INTRODUCTION

In this chapter we shall discuss briefly the nonclassical umbral calculi. It is our intention to give only an introduction to the theory. For a more detailed discussion we refer the reader to Roman [6–8, 10].

Let c_n be a sequence of nonzero constants. If $n!$ is replaced by c_n throughout the preceding theory, then virtually all of the results remain true, mutatis mutandis. In this way each sequence c_n gives rise to a distinct umbral calculus.

Actually, Ward [2] seems to have been the first to suggest such a generalization (of the calculus of finite differences) in 1936, but the idea remained relatively undeveloped until quite recently, perhaps due to a feeling that it was mainly generalization for its own sake. Our purpose here is to indicate that this is not the case.

In the next section we list the principal results. In Section 3 we discuss a particular delta series, which for certain values of c_n gives rise to the Gegenbauer and Chebyshev polynomials. In the final section we discuss a particular nonclassical umbral calculus, which is known as the q-umbral calculus. Sections 3 and 4 are independent of each other.

2. THE PRINCIPAL RESULTS

In this section we consider the effect of replacing $n!$ by c_n in the preceding theory. Since this chapter is intended only to be isagogic, we merely give a summary of the main results. In most cases the proofs follow lines identitical to those of the classical case, and for these we refer the reader to Roman [6–8, 10].

Let c_n be a fixed sequence of nonzero constants. Then we may map the algebra $\mathscr{F}$ of all formal power series in t isomorphically onto the vector space P^* of all linear functionals on P, in a manner completely analogous to that of Section 2.1, by setting

$$\langle t^k \mid x^n \rangle = c_n \delta_{n,k}.$$

Again we obscure this isomorphism and think of $\mathscr{F}$ as the algebra of all linear functionals on P. If

$$f(t) = \sum_{k=0}^{\infty} a_k t^k,$$

then we have

$$\langle f(t) \mid x^n \rangle = c_n a_n.$$

If we define the continuous operator σ_t on $\mathscr{F}$ by

$$\sigma_t t^n = (c_n / c_{n-1}) t^{n-1},$$

then Theorem 2.1.10 has the following analog.

Theorem 6.2.1 If $f(t)$ is in $\mathscr{F}$, then

$$\langle f(t) \mid xp(x) \rangle = \langle \sigma_t f(t) \mid p(x) \rangle$$

for all polynomials $p(x)$.

The evaluation functional is

$$\varepsilon_y(t) = \sum_{k=0}^{\infty} \frac{y^k}{c_k} t^k,$$

which is the analog of the exponential series for the c_n-umbral calculus. As one might expect, this series lies at the heart of the theory. We have

$$\langle \varepsilon_y(t) \mid p(x) \rangle = p(y)$$

for all polynomials $p(x)$.

With regard to linear operators, it is clear that the statement of Theorem 2.2.5,

$$\langle f(t)g(t) \mid p(x) \rangle = \langle f(t) \mid g(t)p(x) \rangle,$$

is all but indispensable. Guided by this requirement, we are forced, quite willingly, to adopt the definition

$$t^k x^n = (c_n / c_{n-k}) x^{n-k}$$

and, more generally,

$$f(t)x^n = \sum_{k=0}^{\infty} a_k t^k x^n = \sum_{k=0}^{n} \frac{c_n}{c_{n-k}} a_k x^{n-k}.$$

The operator $\varepsilon_y(t)$ satisfies

$$\varepsilon_y(t)x^n = \sum_{k=0}^{n} \frac{c_n}{c_k c_{n-k}} y^{n-k} x^k,$$

this sum being the analog of $(x + y)^n$. In fact, following old-style umbral notation, one would write this sum as $(x + y)^n$.

A sequence of polynomials $s_n(x)$ is *Sheffer* for the pair $\big(g(t), f(t)\big)$ if

$$\langle g(t)f(t)^k \,|\, s_n(x)\rangle = c_n \delta_{n,k}$$

for all $n, k \geq 0$. As usual we require that $o(g(t)) = 0$ and $o(f(t)) = 1$. Associated and Appell sequences are defined as expected, that is, with $g(t) = 1$ and $f(t) = t$, respectively. All of the principal results of Section 2.3 have analogs for this definition.

Theorem 6.2.2 (The Expansion Theorem) Let $s_n(x)$ be Sheffer for $\big(g(t), f(t)\big)$. Then for any $h(t)$ in $\mathscr{F}$,

$$h(t) = \sum_{k=0}^{\infty} \frac{\langle h(t) \,|\, s_k(x)\rangle}{c_k} g(t)f(t)^k.$$

Theorem 6.2.3 (The Polynomial Expansion Theorem)   Let $s_n(x)$ be Sheffer for $\big(g(t), f(t)\big)$. Then for any polynomial $p(x)$,

$$p(x) = \sum_{k \geq 0} \frac{\langle g(t)f(t)^k \,|\, p(x)\rangle}{c_k} s_k(x).$$

Theorem 6.2.4   The sequence $s_n(x)$ is Sheffer for $\big(g(t), f(t)\big)$ if and only if

$$\frac{1}{g(\bar{f}(t))} \varepsilon_y(\bar{f}(t)) = \sum_{k=0}^{\infty} \frac{s_k(y)}{c_k} t^k$$

for all constants y in c.

Boas and Buck [1], in their work on polynomial expansions of analytic functions, define a sequence of polynomials to be of generalized Appell type if it has the above generating function, but without the factor $1/c_k$ on the right.

Theorem 6.2.5 (The Conjugate Representation)   The sequence $s_n(x)$ is Sheffer for $\big(g(t), f(t)\big)$ if and only if

$$s_n(x) = \sum_{k=0}^{n} \frac{\langle g(\bar{f}(t))^{-1}\bar{f}(t)^k \,|\, x^n\rangle}{c_k} x^k.$$

Theorem 6.2.6   The sequence $s_n(x)$ is Sheffer for $\big(g(t), f(t)\big)$ if and only if $g(t)s_n(x)$ is the associated sequence for $f(t)$.

Theorem 6.2.7 A sequence $s_n(x)$ is Sheffer for $(g(t), f(t))$, for some invertible $g(t)$, if and only if

$$f(t)s_n(x) = \frac{c_n}{c_{n-1}} s_{n-1}(x).$$

Theorem 6.2.8 (The Sheffer Identity) A sequence $s_n(x)$ is Sheffer for $(g(t), f(t))$, for some invertible $g(t)$, if and only if

$$\varepsilon_y(t)s_n(x) = \sum_{k=0}^{n} \frac{c_n}{c_k c_{n-k}} p_k(y)s_{n-k}(x)$$

for all constants y, where $p_n(x)$ is associated to $f(t)$.

If $p_n(x)$ is the associated sequence for $f(t)$ in the c_n-umbral calculus, then the *transfer operator* for $p_n(x)$, or for $f(t)$, is the operator defined by

$$\lambda: x^n \to p_n(x).$$

In the classical case, a transfer operator is an umbral operator. Sheffer operators are defined in an analogous manner. As before, the adjoint map is a bijection between the set of all transfer operators on P and the set of all automorphisms of $\mathscr{F}$ (Theorem 3.4.2).

The theory of umbral shifts takes on a slight complication in the non-classical case. Let $p_n(x)$ be associated to $f(t)$. In order to preserve the characterization of umbral shifts given in Theorem 3.6.1, we must take the umbral shift $\theta_{f(t)}$ to be

$$\theta_{f(t)}: p_n(x) \to \frac{(n+1)c_n}{c_{n+1}} p_{n+1}(x).$$

Of course, in the classical case, for which $c_n = n!$, we get the classical umbral shift of Section 3.6. Thus θ is the umbral shift for $f(t)$ if and only if θ^* is the derivation $\partial_{f(t)}$ on $\mathscr{F}$. Notice that

$$\theta_t: x^n \to \frac{(n+1)c_n}{c_{n+1}} x^{n+1}$$

is the analog of multiplication by x. It is interesting to observe that

$$\theta_t t x^n = \frac{c_n}{c_{n-1}} \theta_t x^{n-1} = n x^n,$$

and so

$$\theta_t t = xD,$$

where D is the ordinary derivative.

We still have a very useful recurrence formula.

Theorem 6.2.9   (The Recurrence Formula)   If $s_n(x)$ is Sheffer for $(g(t), f(t))$, then

$$(n + 1)s_{n+1}(x) = \frac{c_{n+1}}{c_n}\left(\theta_t - \frac{g'(t)}{g(t)}\right)\frac{1}{f'(t)}s_n(x).$$

Finally, we mention the transfer formulas.

Theorem 6.2.10   If $p_n(x)$ is associated to $f(t)$, then

(i) $p_n(x) = f'(t)\left(\dfrac{f(t)}{t}\right)^{-n-1}x^n$ and

(ii) $p_n(x) = \dfrac{c_n}{nc_{n-1}}\theta_t\left(\dfrac{f(t)}{t}\right)^{-n}x^{n-1}\ (n > 0).$

3. A PARTICULAR DELTA SERIES AND THE GEGENBAUER POLYNOMIALS

In this section we briefly discuss the delta series

$$f(t) = \frac{\sqrt{1 - t^2} - 1}{t}.$$

This will lead us eventually to the Gegenbauer and Chebyshev polynomials. We begin by observing that

$$f(t) = \frac{-t}{1 + \sqrt{1 - t^2}},$$

$$\bar{f}(t) = \frac{-2t}{1 + t^2},$$

$$f'(t) = \frac{f(t)}{t\sqrt{1 - t^2}}.$$

Also, since

$$(1 + \sqrt{1 - z})^{-\alpha} = 2^{-\alpha}\left[1 + \sum_{j=1}^{\infty}\binom{\alpha + 2j - 1}{j - 1}\frac{\alpha}{j4^j}t^j\right] \tag{6.3.1}$$

(see, for example, Rainville [1, p. 70]), we have

$$f(t)^k = (-2)^{-k}\left[t^k + \sum_{j=1}^{\infty}\binom{k + 2j - 1}{j - 1}\frac{k}{j4^j}t^{2j+k}\right]. \tag{6.3.2}$$

The associated sequence $p_n(x)$ for $f(t)$ has the generating function

$$\sum_{k=0}^{\infty} \frac{p_k(x)}{c_k} t^k = \varepsilon_x\left(\frac{-2t}{1+t^2}\right).$$

An explicit expression for $p_n(x)$ comes from the conjugate representation

$$p_n(x) = \sum_{k=0}^{n} \frac{1}{c_k}\left\langle\left(\frac{-2t}{1+t^2}\right)^k \middle| x^n\right\rangle x^k.$$

But

$$\left\langle\left(\frac{-2t}{1+t^2}\right)^k \middle| x^n\right\rangle = (-2)^k \sum_{j=0}^{k}\binom{-k}{j}\langle t^{2j+k} | x^n\rangle$$

$$= (-2)^k \sum_{j=0}^{k}\binom{-k}{j} c_n \delta_{n, 2j+k}$$

$$= \begin{cases} (-2)^{n-2j}\binom{2j-n}{j} c_n, & k = n-2j, \\ 0, & n-k \quad \text{odd}, \end{cases}$$

and so

$$p_n(x) = \sum_{j=0}^{[n/2]}\binom{2j-n}{j}\frac{c_n}{c_{n-2j}}(-2x)^{n-2j}.$$

We shall concentrate mostly on recurrence formulas in this brief discussion, but let us give one example of the polynomial expansion theorem, namely,

$$x^n = \sum_{k=0}^{n}\frac{\langle f(t)^k | x^n\rangle}{c_k}p_k(x),$$

which, from (6.3.2), is

$$x^n = (-2)^{-n}\left[p_n(x) + \sum_{j\geq 1}\binom{n-1}{j-1}\frac{n-2j}{j}\frac{c_n}{c_{n-2j}}p_{n-2j}(x)\right].$$

Let us turn to some recurrence formulas. Theorem 6.2.7 implies that

$$\frac{\sqrt{1-t^2}-1}{t}p_n(x) = \frac{c_n}{c_{n-1}}p_{n-1}(x),$$

which is equivalent to

$$\sqrt{1-t^2}\,p_n(x) = \frac{c_n}{c_{n-1}}tp_{n-1}(x) + p_n(x). \qquad (6.3.3)$$

But also,

$$\frac{-t}{1 + \sqrt{1 - t^2}} p_n(x) = \frac{c_n}{c_{n-1}} p_{n-1}(x),$$

which, for n replaced by $n + 1$, is equivalent to

$$\sqrt{1 - t^2}\, p_n(x) = -\frac{c_n}{c_{n+1}} t p_{n+1}(x) - p_n(x).$$

Equating these two expressions for $\sqrt{1 - t^2}\, p_n(x)$ gives the recurrence

$$\frac{c_n}{c_{n+1}} t p_{n+1}(x) + \frac{c_n}{c_{n-1}} t p_{n-1}(x) + 2 p_n(x) = 0. \tag{6.3.4}$$

We remark that (6.3.4) holds for any Sheffer sequence that uses $f(t)$ as its delta series.

The recurrence formula for $p_n(x)$ is

$$(n + 1) p_{n+1}(x) = \frac{c_n + 1}{c_n} \theta_t \frac{1}{f'(t)} p_n(x)$$

$$= \frac{c_n + 1}{c_n} \theta_t \frac{t \sqrt{1 - t^2}}{f(t)} p_n(x).$$

Writing $p_n(x) = (c_n/c_{n+1}) f(t) p_{n+1}(x)$, substituting into the above equation, and replacing $n + 1$ by n give

$$n p_n(x) = \theta_t t \sqrt{1 - t^2}\, p_n(x).$$

Employing (6.3.3), we obtain

$$n p_n(x) = \theta_t t \left(\frac{c_n}{c_{n-1}} t p_{n-1}(x) + p_n(x) \right),$$

and recalling that $\theta_t t = xD$, we have

$$(xD - n) p_n(x) + \frac{c_n}{c_{n-1}} x D t p_{n-1}(x) = 0. \tag{6.3.5}$$

Equations (6.3.4) and (6.3.5) can be used to derive an operator equation involving $p_n(x)$ only. First, we observe that

$$t \theta_t - \theta_t t = 1.$$

Multiplying by t on the right and replacing $\theta_t t$ by xD, we get

$$t x D - x D t = t. \tag{6.3.6}$$

(Of course, this could easily be verified directly.) Applying xD to (6.3.4), we get

$$\frac{c_n}{c_{n+1}}xDtp_{n+1}(x) + \frac{c_n}{c_{n-1}}xDtp_{n-1}(x) + 2xDp_n(x) = 0.$$

We then solve for $(c_n/c_{n-1})xDtp_{n-1}(x)$ in (6.3.5) and substitute to get

$$\frac{c_n}{c_{n+1}}xDtp_{n+1}(x) + (xD + n)p_n(x) = 0.$$

If $(xD - n)$ is applied to this, we obtain

$$\frac{c_n}{c_{n+1}}xD(xD - n)tp_{n+1}(x) + (xD - n)(xD + n)p_n(x) = 0,$$

and using (6.3.6), we have

$$\frac{c_n}{c_{n+1}}xDt(xD - n - 1)p_{n+1}(x) + (xD - n)(xD + n)p_n(x) = 0.$$

Again, using (6.3.5), but with n replaced by $n + 1$, we get

$$-(xDt)^2 p_n(x) + (xD - n)(xD + n)p_n(x) = 0$$

or

$$[(xDt)^2 - (xD)^2 + n^2]p_n(x) = 0.$$

Finally, since

$$(xDt)^2 = xDtxDt = xD(xDt + t)t = xD(xD + 1)t^2,$$

we have

$$[xD(xD + 1)t^2 - (xD)^2 + n^2]p_n(x) = 0. \tag{6.3.7}$$

Now let us consider the Sheffer sequence $s_n^{(\mu)}(x)$ for the pair

$$g(t) = \left(\frac{2}{1 + \sqrt{1 - t^2}}\right)^\mu,$$

$$f(t) = \frac{\sqrt{1 - t^2} - 1}{t}.$$

In the language of Section 5.3

$$s_n^{(\mu)}(x) = \left(\frac{1 + \sqrt{1 - t^2}}{2}\right)^\mu p_n(x)$$

is a cross sequence.

It is easy to see that

$$g(\bar{f}(t)) = (1 + t^2)^\mu,$$

and so the generating function for $s_n^{(\mu)}(x)$ is

$$\sum_{k=0}^{\infty} \frac{s_k^{(\mu)}(x)}{c_k} t^k = (1 + t^2)^{-\mu} \varepsilon_x \left(\frac{-2t}{1 + t^2} \right). \tag{6.3.8}$$

To obtain the conjugate representation we observe that

$$\langle g(\bar{f}(t))^{-1} \bar{f}(t)^k \mid x^n \rangle = \langle (-2)^k t^k (1 + t^2)^{-k-\mu} \mid x^n \rangle$$

$$= (-2)^k \frac{c_n}{c_{n-k}} \langle (1 + t^2)^{-k-\mu} \mid x^{n-k} \rangle$$

$$= (-2)^k \frac{c_n}{c_{n-k}} \sum_{j=0}^{\infty} \binom{-k-\mu}{j} \langle t^{2j} \mid x^{n-k} \rangle$$

$$= \begin{cases} (-2)^{n-2j} c_n \binom{2j-n-\mu}{j}, & k = n - 2j, \\ 0, & n-k \quad \text{odd}. \end{cases}$$

Hence

$$s_n^{(\mu)}(x) = \sum_{k=0}^{n/2} \binom{2j-n-\mu}{j} \frac{c_n}{c_{n-2j}} (-2x)^{n-2j}.$$

According to (6.3.1),

$$g(t) f(t)^k = 2^\mu (-t)^k (1 + \sqrt{1 - t^2})^{-\mu-k}$$

$$= (-2)^k \left[t^k + \sum_{j=1}^{\infty} \binom{\mu+k+2j-1}{j-1} \frac{\mu+k}{j4^j} t^{2j+k} \right],$$

and, after a few simple manipulations, we arrive at the expansion of x^n in terms of $s_n^{(\mu)}(x)$,

$$(-2)^n x^n = s_n^{(\mu)}(x) + \sum_{j=1}^{[n/2]} \binom{\mu+n-1}{j-1} \frac{c_n}{c_{n-2j}} \frac{\mu+n-2j}{j} s_{n-2j}^{(\mu)}(x).$$

Let us turn to recurrence formulas. We recall that (6.3.4) holds for $s_n^{(\mu)}(x)$,

$$\frac{c_n}{c_{n+1}} t s_{n+1}^{(\mu)}(x) + \frac{c_n}{c_{n-1}} t s_{n-1}^{(\mu)}(x) + 2 s_n^{(\mu)}(x) = 0. \tag{6.3.9}$$

The recurrence formula is

$$(n+1) s_{n+1}^{(\mu)}(x) = \frac{c_n+1}{c_n} \left(\theta_t - \frac{g'(t)}{g(t)} \right) \frac{1}{f'(t)} s_n^{(\mu)}(x).$$

Now

$$\frac{g'(t)}{g(t)} = -\mu t f'(t)$$

and

$$\frac{f(t)}{f'(t)} = t\sqrt{1 - t^2},$$

and so

$$(n + 1)s_{n+1}^{(\mu)}(x) = \frac{c_n + 1}{c_n}\left(\theta_t \frac{1}{f'(t)} + \mu t\right)s_n^{(\mu)}(x)$$

$$= \theta_t \frac{f(t)}{f'(t)} s_{n+1}^{(\mu)}(x) + \frac{c_n + 1}{c_n}\mu t s_n^{(\mu)}(x)$$

$$= \theta_t t \sqrt{1 - t^2}\, s_{n+1}^{(\mu)}(x) + \frac{c_n + 1}{c_n}\mu t s_n^{(\mu)}(x).$$

Replacing $n + 1$ by n and using (6.3.3), which holds for $s_n^{(\mu)}(x)$, we have

$$ns_n^{(\mu)}(x) = \theta_t t s_n^{(\mu)}(x) + \frac{c_n}{c_{n-1}}\mu t s_{n-1}^{(\mu)}(x)$$

$$= xD\left(\frac{c_n}{c_{n-1}} t s_{n-1}^{(\mu)}(x) + s_n^{(\mu)}(x)\right) + \frac{c_n}{c_{n-1}}\mu t s_{n-1}^{(\mu)}(x),$$

which is

$$(xD - n)s_n^{(\mu)}(x) + \frac{c_n}{c_{n-1}}(xD + \mu)t s_{n-1}^{(\mu)}(x) = 0. \qquad (6.3.10)$$

Again, we can use (6.3.9) and (6.3.10) to obtain an operator equation for $s_n^{(\mu)}(x)$. First, we apply $(xD + \mu)$ to (6.3.9),

$$\frac{c_n}{c_{n+1}}(xD + \mu)t s_{n+1}^{(\mu)}(x) + \frac{c_n}{c_{n-1}}(xD + \mu)t s_{n-1}^{(\mu)}(x) + 2(xD + \mu)s_n^{(\mu)}(x) = 0.$$

Substituting into (6.3.10), we have, after collecting like terms,

$$\frac{c_n}{c_{n+1}}(xD + \mu)t s_{n+1}^{(\mu)}(x) + (xD + n + 2\mu)s_n^{(\mu)}(x) = 0.$$

Applying $(xD - n)$ to this and using $(xD - n)t = t(xD - n - 1)$, we get

$$\frac{c_n}{c_{n+1}}(xD + \mu)t(xD - n - 1)s_{n+1}^{(\mu)}(x) + (xD - n)(xD + n + 2\mu)s_n^{(\mu)}(x) = 0.$$

Using (6.3.10), with n replaced by $n + 1$, gives

$$[((xD + \mu)t)^2 - (xD - n)(xD + n + 2\mu)]s_n^{(\mu)}(x) = 0.$$

Finally, from (6.3.6) we have

$$[(xD + \mu)(xD + \mu + 1)t^2 - (xD - n)(xD + n + 2\mu)]s_n^{(\mu)}(x) = 0. \quad (6.3.11)$$

Let us now specialize the sequence c_n. The case $c_n = n!$ is the classical one, and we shall not discuss it except to remind the reader of the example given in 4.12, at the end of Section 4 of Chapter 4. Let us take

$$c_n = 1 \bigg/ \binom{-\lambda}{n},$$

where λ is not a negative integer. Then

$$\frac{c_n}{c_{n-1}} = \frac{n}{-\lambda - n + 1}$$

and

$$tx^n = \frac{n}{-\lambda - n + 1} x^{n-1},$$

from which it follows that

$$t = -(\lambda + xD)^{-1}D.$$

Actually, the motivation for choosing c_n as we did is that

$$\varepsilon_y(t) = \sum_{k=0}^{\infty} \binom{-\lambda}{k} y^k t^k = (1 + yt)^{-\lambda}.$$

We consider first $p_n(x)$. The generating function is

$$\sum_{k=0}^{\infty} \binom{-\lambda}{k} p_k(x)t^k = \left(1 - \frac{2xt}{1 + t^2}\right)^{-\lambda}$$

$$= (1 + t^2)^\lambda (1 - 2xt + t^2)^{-\lambda}.$$

(The reader may begin to notice a Gegenbauer flavor here.)

The conjugate representation is

$$\binom{-\lambda}{n} p_n(x) = \sum_{j=0}^{[n/2]} \binom{-\lambda}{n - 2j}\binom{2j - n}{j}(-2x)^{n-2j}.$$

Dividing (6.3.4) by c_n gives

$$\binom{-\lambda}{n + 1} t p_{n+1}(x) + \binom{-\lambda}{n - 1} t p_{n-1}(x) + 2\binom{-\lambda}{n} p_n(x) = 0.$$

Since $-(\lambda + xD)t = D$, we get

$$\binom{-\lambda}{n+1}Dp_{n+1}(x) + \binom{-\lambda}{n-1}Dp_{n-1}(x) - 2\binom{-\lambda}{n}(\lambda + xD)p_n(x) = 0, \quad (6.3.12)$$

which holds also for any Sheffer sequence using $f(t)$ as its delta series.

Equation (6.3.5) is, after dividing by c_n and applying $-(\lambda + xD)$,

$$\binom{-\lambda}{n}(xD + \lambda)(xD - n)p_n(x) - \binom{-\lambda}{n-1}xD^2p_{n-1}(x) = 0.$$

Finally, since $D(\lambda + xD)^{-1} = (\lambda + 1 + xD)^{-1}D$, Eq. (6.3.7) is

$$[xD(xD + 1)(xD + \lambda)^{-1}(xD + \lambda + 1)^{-1}D^2 - (xD)^2 + n^2]p_n(x) = 0.$$

Next, let us consider the sequence $s_n^{(\mu)}(x)$ for the above choice of c_n. The generating function is (6.3.8),

$$\sum_{k=0}^{\infty}\binom{-\lambda}{k}s_k^{(\mu)}(x)t^k = (1 + t^2)^{\lambda - \mu}(1 - 2xt + t^2)^{-\lambda}.$$

Thus if $\mu = \lambda$, the polynomials

$$c_n^{(\lambda)}(x) = \binom{-\lambda}{n}s_n^{(\lambda)}(x)$$

are the Gegenbauer polynomials (Erdelyi [1, Vol. 2, p. 177]). Taking $\mu = \lambda = 1$, we get the Chebyshev polynomials of the second kind. The Chebyshev polynomials of the first kind are Sheffer for a different $g(t)$.

The conjugate representation is

$$\binom{-\lambda}{n}s_n^{(\mu)}(x) = \sum_{k=0}^{[n/2]}\binom{2j - n - \mu}{j}\binom{-\lambda}{n - 2j}(-2x)^{n - 2j},$$

and for $\mu = \lambda$ we get

$$C_n^{(\lambda)}(x) = \sum_{k=0}^{[n/2]}\frac{(-1)^k\lambda^{(n-j)}}{j!(n - 2j)!}(2x)^{n - 2j},$$

which is a well-known expression for the Gegenbauer polynomials (Erdelyi [1, Vol. 2, p. 175]).

Our example of the polynomial expansion theorem is

$$\binom{-\lambda}{n}(-2x)^n = s_n^{(\mu)}(x) + \sum_{j=1}^{[n/2]}\binom{\mu + n - 1}{j - 1}\binom{-\lambda}{n - 2j}\frac{\mu + n - 2j}{j}s_{n-2j}^{(\mu)}(x).$$

Let us turn to the recurrence formulas. Equation (6.3.9) is

$$\binom{-\lambda}{n+1}Ds_{n+1}^{(\mu)}(x) + \binom{-\lambda}{n-1}Ds_{n-1}^{(\mu)}(x) - 2\binom{-\lambda}{n}(\lambda + xD)s_n^{(\mu)}(x) = 0.$$

When $\mu = \lambda$, we get

$$DC_{n+1}^{(\lambda)}(x) + DC_{n-1}^{(\lambda)}(x) - 2(\lambda + xD)C_n^{(\lambda)}(x) = 0,$$

which is also a well-known formula for the Gegenbauer polynomials (Erdelyi [1, Vol. 2, p. 176]).

Equation (6.3.10) is

$$\binom{-\lambda}{n}(xD + \lambda)(xD - n)s_n^{(\mu)}(x) - \binom{-\lambda}{n-1}(xD + \mu)Ds_{n-1}^{(\mu)}(x) = 0.$$

In the Gegenbauer case, since $\mu = \lambda$, we may apply $(xD + \lambda)^{-1}$ to get

$$(n - xD)C_n^{(\lambda)}(x) + DC_{n-1}^{(\lambda)}(x) = 0.$$

Finally, Eq. (6.3.11) is

$$[(xD + \mu)(xD + \mu + 1)(xD + \lambda)^{-1}D(xD + \lambda)^{-1}D$$
$$- (xD - n)(xD + n - 2\mu)]s_n^{(\mu)}(x) = 0,$$

and since $D(xD + \lambda)^{-1} = (xD + \lambda + 1)^{-1}D$, we get

$$[(xD + \mu)(xD + \mu + 1)(xD + \lambda)^{-1}(xD + \lambda + 1)^{-1}D^2$$
$$- (xD - n)(xD + n - 2\mu)]s_n^{(\mu)}(x) = 0.$$

In the Gegenbauer case, when $\mu = \lambda$, we get

$$[D^2 - (xD - n)(xD + n - 2\lambda)]C_n^{(\lambda)}(x) = 0$$

or, using $(xD)^2 = x^2D^2 + xD$,

$$[(1 - x^2)D^2 - (1 + 2\lambda)D + n(n + 2\lambda)]C_n^{(\lambda)}(x) = 0,$$

which is the well-known differential equation for the Gegenbauer polynomials.

In this section we have shown that the Gegenbauer polynomials fit nicely into the umbral catalog. There are indeed other important polynomial sequences that share this property, but further discussion would be inappropriate here. For a short discussion of the Jacobi polynomials we refer the reader to Roman [6].

4. THE q-UMBRAL CALCULUS

One of the most interesting of the nonclassical umbral calculi is defined by setting

$$c_n = \frac{(1 - q)(1 - q^2)\cdots(1 - q^n)}{(1 - q)^n},$$

where $q \neq 1$. This is the *q-umbral calculus*. We have

$$\frac{c_n}{c_{n-1}} = \frac{1 - q^n}{1 - q}$$

and

$$tx^n = \frac{1 - q^n}{1 - q} x^{n-1} = \frac{x^n - (qx)^n}{x - qx},$$

and so

$$tp(x) = \frac{p(x) - p(qx)}{x - qx}.$$

The operator t is known as the *q-derivative*, although some use this term for the operator $(1 - q)t$.

The *q-binomial coefficient* is

$$\binom{n}{k}_q = \frac{c_n}{c_k c_{n-k}}$$

$$= \frac{(1 - q) \cdots (1 - q^n)}{(1 - q) \cdots (1 - q^k)(1 - q) \cdots (1 - q^{n-k})}.$$

This is also known as the Gaussian coefficient and has important combinatorial significance. In Goldman and Rota [1] it is shown that when q is a prime power, $\binom{n}{k}_q$ is the number of vector subspaces of dimension k of a vector space of dimension n over the finite field $GF[q]$.

We also have

$$\varepsilon_y(t) = \sum_{k=0}^{\infty} \frac{((1 - q) yt)^k}{(1 - q) \cdots (1 - q^k)}$$

and

$$\varepsilon_y(t)x^n = \sum_{k=0}^{n} \binom{n}{k}_q y^k x^{n-k}.$$

It is very easy to see that

$$y\varepsilon_y(t) = \frac{\varepsilon_y(t) - \varepsilon_y(qt)}{t - qt}$$

or

$$\varepsilon_y(qt) = \left(1 - (1 - q) yt\right)\varepsilon_y(t).$$

Recalling that $\sigma_t \colon t^n \to (c_n/c_{n-1})t^{n-1}$, we have

$$\sigma_t f(t) = \frac{f(t) - f(qt)}{t - qt}.$$

The q-derivative operator t, and so also σ_t, satisfies a q-Leibniz formula

$$\sigma_t^n\big(f(t)g(t)\big) = \sum_{k=0}^{n}\binom{n}{k}_q q^{-k(n-k)}\sigma_t^k f(t)\sigma_t^{n-k}g(q^k t).$$

For an umbral proof of this see Roman [6].

As Andrews [1] has observed, the series $\varepsilon_y(t)$ can be expressed as an infinite product

$$\varepsilon_1\!\left(\frac{t}{1-q}\right) = \prod_{j=0}^{\infty}\frac{1}{1-q^j t}$$

that is valid for $|q| < 1$ and $|t| < 1/(1-q)$.

Many of the combinatorial properties of the q-binomial coefficients can be derived easily by umbral means. We observe first that

$$\binom{n}{k}_q = \frac{c_n}{c_k c_{n-k}}$$

$$= \frac{1}{c_k}\langle\varepsilon_1(t)\mid t^k x^n\rangle$$

$$= \left\langle\varepsilon_1(t)\frac{t^k}{c_k}\;\middle|\;x^n\right\rangle.$$

From this we obtain some of the simplest properties of $\binom{n}{k}_q$, for example,

$$\binom{n}{k}_q = \left\langle\varepsilon_1(t)\frac{t^k}{c_k}\;\middle|\;x^n\right\rangle$$

$$= \frac{c_{k-1}}{c_k}\left\langle\varepsilon_1(t)\frac{t^{k-1}}{c_{k-1}}\;\middle|\;tx^n\right\rangle$$

$$= \frac{c_{k-1}}{c_k}\frac{c_n}{c_{n-1}}\binom{n-1}{k-1}_q = \frac{1-q^n}{1-q^k}\binom{n-1}{k-1}_q,$$

and, by using the q-Leibniz formula,

$$\binom{n}{k}_q = \left\langle\varepsilon_1(t)\frac{t^k}{c_k}\;\middle|\;x^n\right\rangle$$

$$= \left\langle\sigma_t\varepsilon_1(t)\frac{t^k}{c_k}\;\middle|\;x^{n-1}\right\rangle$$

$$= \left\langle\varepsilon_1(t)\frac{t^{k-1}}{c_{k-1}} + \varepsilon_1(t)\frac{q^k t^k}{c_k}\;\middle|\;x^{n-1}\right\rangle$$

$$= \binom{n-1}{k-1}_q + q^k\binom{n-1}{k}_q.$$

Let us give a somewhat more delicate example. It is easy to see, by using the q-Leibniz formula, that

$$\sigma_t \varepsilon_y(t) = y \varepsilon_y(t)$$

and

$$\sigma_t \varepsilon_y(t) \varepsilon_z(t) = \left(y + z - (1 - q) yzt \right) \varepsilon_y(t) \varepsilon_z(t).$$

Then

$$\sum_{k=0}^{n} \binom{n}{k}_q y^k z^{n-k} = \left\langle \varepsilon_z(t) \middle| \sum_{k=0}^{n} \binom{n}{k}_q y^k x^{n-k} \right\rangle$$

$$= \left\langle \varepsilon_z(t) \middle| \varepsilon_y(t) x^n \right\rangle$$

$$= \left\langle \varepsilon_y(t) \varepsilon_z(t) \middle| x^n \right\rangle$$

$$= \left\langle \sigma_t \varepsilon_y(t) \varepsilon_z(t) \middle| x^{n-1} \right\rangle$$

$$= \left\langle \left(y + z - (1 - q) yzt \right) \varepsilon_y(t) \varepsilon_z(t) \middle| x^{n-1} \right\rangle$$

$$= (y + z) \sum_{k=0}^{n-1} \binom{n-1}{k}_q y^k z^{n-1-k}$$

$$- (1 - q^{n-1}) yz \sum_{k=0}^{n-2} \binom{n-2}{k}_q y^k z^{n-2-k}.$$

Now if we take $y = z = 1$, then

$$\sum_{k=0}^{m} \binom{m}{k}_q y^k z^{m-k} = \sum_{k=0}^{m} \binom{m}{k}_q$$

is the total number of subspaces, of all dimensions, of a vector space of dimension m over the field $GF[q]$. Denoting this by G_m, we have the recurrence formula

$$G_n = 2G_{n-1} - (1 - q^{n-1}) G_{n-2}.$$

Taking $y = -1$ and $z = 1$, we obtain, for the alternating sum

$$A_m = \sum_{k=0}^{m} (-1)^k \binom{m}{k}_q,$$

the recurrence

$$A_n = (1 - q^{n-1}) A_{n-2}.$$

Combinatorial considerations aside, the q-umbral calculus, and other related umbral calculi, plays a definite role in the theory of basic hypergeometric series and hence in the theory of numbers. However, this is not the place to enter into a general discussion, and we refer the reader to Slater [1] and to Hardy and Wright [1] for the q-analog of the hypergeometric series.

We shall content ourselves with a brief discussion of two important Sheffer sequences, namely, the Appell sequences for $\varepsilon_y(t)$ and $\varepsilon_y(t)^{-1}$.

Let us use the notation

$$s_n(x) = [x]_{y,n}$$

for the Appell sequence for

$$g(t) = \varepsilon_y(t).$$

In order to determine $[x]_{y,n}$ we first observe that

$$[y]_{y,n+1} = \langle \varepsilon_y(t) \,|\, [x]_{y,n+1} \rangle$$
$$= c_{n+1}\delta_{n+1,0}$$
$$= 0,$$

and so we may set

$$[x]_{y,n+1} = (x - y)r_n(x).$$

The q-Leibniz formula gives

$$c_{k+1}\delta_{n+1,k+1} = \langle \varepsilon_y(t)t^{k+1} \,|\, [x]_{y,n+1} \rangle$$
$$= \langle \varepsilon_y(t)t^{k+1} \,|\, (x - y)r_n(x) \rangle$$
$$= \langle (\sigma_t - y)\varepsilon_y(t)t^{k+1} \,|\, r_n(x) \rangle$$
$$= \frac{c_{k+1}}{c_k} \langle \varepsilon_y(qt)t^k \,|\, r_n(x) \rangle,$$

and since $\varepsilon_y(qt) = \varepsilon_{qy}(t)$, we get

$$\langle \varepsilon_{qy}(t)t^k \,|\, r_n(x) \rangle = c_k\delta_{n,k}.$$

Thus $r_n(x) = [x]_{qy,n}$ and

$$[x]_{y,n+1} = (x - y)[x]_{qy,n},$$

which leads to

$$[x]_{y,n} = (x - y)(x - qy) \cdots (x - q^{n-1}y).$$

The generating function for $[x]_{y,n}$ is

$$\sum_{k=0}^{\infty} \frac{[x]_{y,k}}{c_k} t^k = \frac{\varepsilon_x(t)}{\varepsilon_y(t)}.$$

Setting $x = 0$ and observing that

$$[0]_{y,k} = (-y)^k q^{\binom{k}{2}},$$

we have

$$\frac{1}{\varepsilon_y(t)} = \sum_{k=0}^{\infty} \frac{(-y)^k}{c_k} q^{\binom{k}{2}} t^k.$$

This gives an explicit expression for $[x]_{y,n}$,

$$[x]_{y,n} = \frac{1}{\varepsilon_y(t)} x^n$$

$$= \sum_{k=0}^{n} \frac{(-y)^k}{c_k} q^{\binom{k}{2}} t^k x^n$$

$$= \sum_{k=0}^{n} \binom{n}{k}_q (-y)^{n-k} q^{\binom{n-k}{2}} x^k.$$

Incidentally, in view of the expression

$$\varepsilon_z\left(\frac{t}{1-q}\right) = \varepsilon_1\left(\frac{zt}{1-q}\right)$$

$$= \prod_{j=0}^{\infty} \frac{1}{1-q^j zt},$$

which holds for $|q| < 1$ and $|t| < 1/(1-q)$, the generating function becomes, after replacing t by $t/(1-q)$,

$$\sum_{k=0}^{\infty} \frac{[x]_{y,k}}{c_k} t^k = \prod_{j=0}^{\infty} \frac{1-q^j yt}{1-q^j zt}.$$

Using the notation of basic hypergeometric series (Slater [1]),

$$c_n(1-q)^n = (q;q)_n$$

and

$$[x]_{y,n} = x^n\left(\frac{y}{x};q\right)_n,$$

this becomes

$$\prod_{j=0}^{\infty} \frac{1-q^j yt}{1-q^j zt} = \sum_{k=0}^{\infty} \frac{(y/x;q)_k}{(q;q)_k} (xt)^k$$

$$= {}_1\phi_0[y/x;q;xt],$$

which is Heine's theorem (Slater [1, p. 92]).

Since $[x]_{y,n}$ is Appell, we have

$$t[x]_{y,n} = \frac{c_n}{c_{n-1}} [x]_{y,n-1},$$

and since t is the q-derivative, we obtain

$$\frac{[x]_{y,n} - [qx]_{y,n}}{x - qx} = \frac{c_n}{c_{n-1}}[x]_{y,n-1}$$

or

$$[x]_{y,n} - [qx]_{y,n} = (1 - q^n)x[x]_{y,n-1}.$$

From the polynomial expansion theorem we get

$$x^n = \sum_{k=0}^{n} \frac{\langle \varepsilon_y(t)t^k \mid x^n \rangle}{c_k}[x]_{y,k}$$

$$= \sum_{k=0}^{n} \binom{n}{k}_q y^{n-k}[x]_{y,k}.$$

The Appell sequence for

$$g(t) = \varepsilon_y(t)^{-1}$$

is

$$H_n(x;y) = \varepsilon_y(t)x^n = \sum_{k=0}^{n} \binom{n}{k}_q y^{n-k}x^k,$$

which is the q-analog of $(x + y)^n$. The polynomials $H_n(x;1)$ are known, however, as the q-Hermite polynomials. (One reason for this is that the polynomials $H_n(x;y)$ have some algebraic properties that are analogous to those of the Hermite polynomials; for example,

$$H_n(x;1)H_m(x;1) = \sum_{k=0}^{n} \binom{n}{k}_q \binom{m}{k}_q c_k(1 - q)^k x^k H_{n+m-2k}(x;1);$$

for another reason see Allaway [2].)

The generating function for the q-Hermite polynomial is

$$\sum_{k=0}^{\infty} \frac{H_k(x;y)}{c_k}t^k = \varepsilon_y(t)\varepsilon_x(t),$$

and if $|q| < 1$ and $|t| < 1/(1 - q)$,

$$\sum_{k=0}^{\infty} \frac{H_k(x;y)}{c_k}\left(\frac{t}{1-q}\right)^k = \prod_{j=0}^{\infty} \frac{1}{(1 - q^j yt)(1 - q^j xt)}.$$

The fact that $H_n(x;y)$ is inverse to $[x]_{y,n}$ under umbral composition can be expressed as

$$x^n = \sum_{k=0}^{n} \binom{n}{k}_q (-y)^{n-k}q^{\binom{n-k}{2}}H_k(x;y).$$

Since $H_n(x; y)$ is Appell, we have

$$tH_n(x; y) = \frac{c_n}{c_{n-1}} H_{n-1}(x; y),$$

which is equivalent to

$$H_n(x; y) - H_n(qx; y) = (1 - q^n)xH_{n-1}(x; y).$$

Finally, from the polynomial expansion theorem we have

$$xH_n(x; y) = \sum_{k=0}^{n+1} \frac{\langle \varepsilon_y(t)^{-1} t^k \mid xH_n(x; y) \rangle}{c_k} H_k(x; y).$$

But by the q-Leibniz formula

$$\langle \varepsilon_y(t)^{-1} t^k \mid xH_n(x; y) \rangle = \langle \sigma_t \varepsilon_y(t)^{-1} t^k \mid H_n(x; y) \rangle$$

$$= \left\langle \frac{c_k}{c_{k-1}} \varepsilon_y(t)^{-1} t^{k-1} - (qt)^k y \varepsilon_y(qt)^{-1} \mid H_n(x; y) \right\rangle$$

$$= c_k \delta_{k, n+1} - y \langle \varepsilon_y(t)^{-1} t^k \mid H_n(qx; y) \rangle,$$

and so

$$xH_n(x; y) = \sum_{k=0}^{n+1} \frac{1}{c_k} (c_k \delta_{k, n+1} - y \langle \varepsilon_y(t)^{-1} t^k \mid H_n(qx; y) \rangle) H_k(x; y)$$

$$= H_{n+1}(x; y) - y \sum_{k=0}^{n+1} \frac{1}{c_k} \langle \varepsilon_y(t)^{-1} t^k \mid H_n(qx; y) \rangle H_k(x; y)$$

$$= H_{n+1}(x; y) - yH_n(qx; y)$$

or

$$H_{n+1}(x; y) = xH_n(x; y) + yH_n(qx; y).$$

We have scratched only the surface of the q-umbral calculus in briefly discussing these two Appell sequences. For a somewhat more detailed discussion along these lines we refer the reader to Roman [6, 10].

There have been several works on the subject of the q-umbral calculus in the past decade. Andrews [1] was the first to make a detailed study. However, his theory differs from the one we have just outlined. For example, the role of the Sheffer identity

$$\varepsilon_y(t)s_n(x) = \sum_{k=0}^{n} \binom{n}{k}_q s_k(x)p_{n-k}(y)$$

is taken by the identity

$$s_n(xy) = \sum_{k=0}^{n} \binom{n}{k}_q s_k(x)y^k p_{n-k}(y).$$

Without going into any detail, it happens that Andrews' theory is actually the theory of Appell sequences in a whole class of umbral calculi, all of which are related to the q-umbral calculus. This has opened up a study of the relationship among the various related umbral calculi. For example, the umbral calculus defined by setting

$$c_n = q^{-\binom{n}{2}} \frac{(1-q)\cdots(1-q^n)}{(1-q)^n}$$

or, more generally,

$$c_n = \frac{1}{[a]_{y,n}} \frac{(1-q)\cdots(1-q^n)}{(1-q)^n}$$

has strong ties to the q-umbral calculus.

The study of these relationships has barely begun and promises to be quite fruitful (Roman [10]).

REFERENCES

This list contains those references used in the text as well as a few others. It is intended only as a guide to further study of the umbral calculus and not as a bibliography on Sheffer sequences.

Allaway, W. R.
  [1]   Extensions of Sheffer polynomial sets, *SIAM J. Math. Anal.* **10** (1979), 38–48.
  [2]   Some properties of the q-Hermite polynomials, *Canad. J. Math.* **32** (1980), 686–694.
  [3]   A comparison of two umbral algebras, *J. Math. Anal. Appl.* **85** (1982), 197–235.

Allaway, W. R., and Yuen, K. W.
  [1]   Ring isomorphisms for the family of Eulerian differential operators, *J. Math. Anal. Appl.* **77** (1980), 245–263.

Al-Salam, W. A.
  [1] q-Appell polynomials, *Ann. Mat. Pura Appl.* **57** (1967), 31–46.

Al-Salam, W. A., and Ismail, M. E. H.
  [1]   Some operational formulas, *J. Math. Anal. Appl.* **51** (1975), 208–218.

Al-Salam, W. A., and Verma, A.
  [1]   Generalized Sheffer polynomials, *Duke Math. J.* **37** (1970), 361–365.

Andrews, G. E.
  [1]   On the foundations of combinatorial theory V, Eulerian differential operators, *Stud. Appl. Math.*, **50** (1971), 345–375.

Bateman, H.
  [1]   The polynomial of Mittag-Leffler, *Proc. Nat. Acad. Sci. U.S.A.* **26** (1940), 491–496.

Bell, E. T.
  [1]   Postulational bases for the umbral calculus, *Amer. J. Math.* **62** (1940), 717–724.

Boas, R. P., and Buck, R. C.
  [1]   "Polynomial Expansions of Analytic Functions," Academic Press, New York, 1964.

Boole, G.
  [1]   "Calculus of Finite Differences," Chelsea, Bronx, New York, 1970.

Brown, J. W.
  [1]   A note on generalized Appell polynomials, *Amer. Math. Monthly* **75** (1968).
  [2]   Generalized Appell connection sequences II, *J. Math. Anal. Appl.* **50** (1975), 458–464.

[3]   On orthogonal sheffer sequences, *Glas. Mat. Ser. III* **10**, 30 (1975), 63–67.

[4]   Steffensen sequences satisfying a certain composition law, *J. Math. Anal. Appl.* **81** (1981), 48–62.

Brown, J. W., and Goldberg, J. L.

[1]   Generalized Appell connection sequences, *J. Math. Anal. Appl.* **46** (1974), 242–248.

Brown, J. W., and Kuezma, M.

[1]   Self-inverse Sheffer sequences, *SIAM J. Math. Anal.* **7** (1976), 723–728.

Brown, J. W., and Roman, S.

[1]   Inverse relations for certain Sheffer sequences, *SIAM J. Math. Anal.* **12** (1981), 186–195.

Brown, R. B.

[1]   Sequences of functions of binomial type, *Discrete Math.* **6** (1973), 313–331.

Burchnall, J. L.

[1]   The Bessel polynomials, *Canad. J. Math.* **3** (1951), 62–68.

Carlitz, L.

[1]   Some polynomials related to theta functions, *Ann. Mat. Pura Appl.* (4) **41** (1955), 359–373.

[2]   A note on the Bessel polynomials, *Duke Math. J.* **24** (1957), 151–162.

Chak, A. M.

[1]   An extension of a class of polynomials I, *Riv. Mat. Univ. Parma* (2) **12** (1971), 47–55.

Chak, A. M., and Agarwal, A. K.

[1]   An extension of a class of polynomials II, *SIAM J. Math. Anal.* **2** (1971), 352–355.

Chak, A. M., and Srivastava, H. M.

[1]   An extension of a class of polynomials III, *Riv. Mat. Univ. Parma* (3) **2** (1973), 11–18.

Chihara, T. S.

[1]   "An Introduction to Orthogonal Polynomials," Gordon & Breach, New York, 1978.

Cigler, J.

[1]   Some remarks on Rota's umbral calculus, *Indag. Math.* **40** (1978), 27–42.

[2]   Operatormethoden für q-Identitäten, *Monatsh. Math.* **88** (1979), 87–105.

[3]   Operatormethoden für q-Identitäten II: q-Laguerre-Polynome, *Monatsh. Math.* **91** (1981), 105–117.

[4]   Operatormethoden für q-Identitäten III: Umbrale Inversion und die Lagrangesche Formal, *Arch. Math.* **35** (1980), 533–543.

Comtet, L.

[1]   "Advanced Combinatorics," Reidel, Boston, 1974.

Erdelyi, A. (ed.)

[1]   "Higher Transcendental Functions," The Bateman Manuscript Project, Vols. I–III, McGraw-Hill, New York, 1953.

Fillmore, J. P., and Williamson, S. G.

[1]   A linear algebra setting for the Mullin–Rota theory of polynomials of binomial type, *Linear and Multilinear Algebra* **1** (1973), 67–80.

Freeman, J. M.

[1]   New solutions to the Rota/Mullin problem of connection constants, *Proc. Southeastern Conf. Combinatorics Graph Theory Comput., 9th, Boca Raton, 1978,* pp. 301–305.

Garsia, A.

[1]   An exposé of the Mullin–Rota theory of polynomials of binomial type, *Linear and Multilinear Algebra* **1** (1973), 47–65.

Garsia, A., and Joni, S. A.

[1]   A new expression for umbral operators and power series inversion, *Proc. Amer. Math. Soc.* **64** (1977), 179–185.

Goldman, J., and Rota, G.-C.
[1] The number of subspaces of a vector space, in "Recent Progress in Combinatorics" (W. T. Tutte, ed.), Academic Press, New York, 1969.
[2] On the foundations of combinatorial theory IV: Finite vector spaces and Eulerian generating functions, *Stud. Appl. Math.* **49**, (1970), 239–258.

Gould, H. W.
[1] A series transformation for finding convolution identities, *Duke Math. J.* **28** (1961), 193–202.
[2] A new convolution formula and some new orthogonal relations for inversion of series, *Duke Math. J.* **29** (1962), 393–404.

Guinand, A.
[1] The umbral method: A survey of elementary mnemonic and manipulative uses, *Amer. Math. Monthly* **86** (1979), 187–195.

Hardy, G. H., and Wright, E. M.
[1] "An Introduction to the Theory of Numbers," Oxford Univ. Press, London and New York, 1979.

Henrici, P.
[1] An algebraic proof of the Lagrange–Burmann formula, *J. Math. Anal. Appl.* **8** (1964), 218–224.

Ihrig, E. C., and Ismail, M. E. H.
[1] On an umbral calculus, *Proc. Southeastern Conf. Combinatorics, Graph Theory Comput., 10th, Boca Raton,* 1979, pp. 523–528.
[2] A q-umbral calculus, *J. Math. Anal. Appl.* **84** (1981), 178–207.

Ismail, M. E. H.
[1] Polynomials of binomial type and approximation theory, *J. Approx. Theory* **23** (1978), 177–186.

Joni, S. A.
[1] Lagrange inversion in higher dimensions and umbral operators, *Linear and Multilinear Algebra* **6** (1978), 111–121.
[2] Expansion formulas I: A general method, *J. Math. Anal. Appl.* **81** (1981), 364–377.
[3] Expansion formulas II: Variations on a theme, *J. Math. Anal. Appl.* **82** (1981), 1–13.

Joni, S. A., and Rota, G.-C.
[1] "Umbral Calculus and Hopf Algebras," Amer. Math. Soc., Providence, Rhode Island, 1978.

Jordan, C.
[1] "Calculus of Finite Differences," Chelsea, Bronx, New York, 1965.

Krall, H. L., and Frink, O.
[1] A new class of orthogonal polynomials: The Bessel polynomials, *Trans. Amer. Math. Soc.* **65** (1948), 100–115.

Krouse, D., and Olive, G.
[1] Binomial functions with the Stirling property, *J. Math. Anal. Appl.* **83** (1981), 110–126.

Labelle, G.
[1] Sur l'inversion et l'iteration continue de series formelles, *European J. Combin.* **1** (2), (1980), 113–138.

Meixner, J.
[1] Orthogonale Polynomsystem mit linern besonderen Gestalt der eryengenden Funktion, *J. London Math. Soc.* **9** (1934), 6–13.

Milne-Thomson, L. M.
[1] "The Calculus of Finite Differences," Chelsea, Bronx, New York, 1960.

Morris, R. A.
[1] Frobenius endomorphisms in the umbral calculus, *Stud. Appl. Math.* **62** (1980), 85–92.

Mullin, R., and Rota, G.-C.
[1] On the foundations of combinatorial theory III: Theory of binomial enumerations, *in* "Graph Theory and Its Applications" (B. Harris, ed.), Academic Press, New York, 1970.

Niederhausen, H.
[1] Sheffer polynomials for computing exact Kolmogorov–Smirnov and Renyi type distributions, Tech. Rep. No. 6, M.I.T., Cambridge, Massachusetts, 1979.

Niven, I.
[1] Formal power series, *Amer. Math. Monthly* **76** (1969), 871–889.

Olive, G.
[1] The b-transform, *J. Math. Anal. Appl.* **60** (1977), 755–778.
[2] Binomial functions and combinatorial mathematics, *J. Math. Anal. Appl.* **70** (1979), 460–473.
[3] A combinatorial approach to generalized powers, *J. Math. Anal. Appl.* **74** (1980), 270–285.

Rainville, E.
[1] "Special functions," Chelsea, Bronx, New York, 1960.

Reiner, D. L.
[1] Sequences of polynomials of fractional binomial type, *Linear and Multilinear Algebra* **5** (1977), 175–179.
[2] The combinatorics of polynomial sequences, *Stud. Appl. Math.* **58** (1978), 95–117.
[3] Eulerian binomial type revisited, *Stud. Appl. Math.* **64** (1981), 89–93.

Riordan, J.
[1] Inverse relations and combinatorial identities, *Amer. Math. Monthly* **71** (1964), 485–498.
[2] "Combinatorial Identities," Wiley, New York, 1968.
[3] "An Introduction to Combinatorial Analysis," Princeton Univ. Press, Princeton, New Jersey, 1980.

Roman, S.
[1] The algebra of formal series, *Adv. in Math.* **31** (1979), 309–339; Erratum **35** (1980), 274.
[2] The algebra of formal series II: Sheffer sequences, *J. Math. Anal. Appl.* **74** (1980), 120–143.
[3] The algebra of formal series III: Several variables, *J. Approx. Theory* **26** (1979), 340–381.
[4] The formula of Faá di Bruno, *Amer. Math. Monthly* **87** (1980), 805–809.
[5] Polynomials, power series and interpolation, *J. Math. Anal. Appl.* **80** (1981), 333–371.
[6] The theory of the umbral calculus I, *J. Math. Anal. Appl.* **87** (1982), 58–115.
[7] The theory of the umbral calculus II, *J. Math. Anal. Appl.* **89** (1982), 290–314.
[8] The theory of the umbral calculus III, *J. Math. Anal. Appl.*, to appear.
[9] Operational formulas, *Linear and Multilinear Algebra* **12** (1982), 1–20.
[10] More on the umbral calculus, with emphasis on the q-umbral calculus, *J. Math. Anal. Appl.*, to appear.

Roman, S., and Rota, G.-C.
[1] The umbral calculus, *Adv. in Math.* **27** (1978), 95–188.

Rota, G.-C.
[1] "Finite Operator Calculus," Academic Press, New York, 1975.

Rota, G.-C., Kahaner, D., and Odlyzko, A.
[1] On the foundations of combinatorial theory VIII: Finite operator calculus, *J. Math. Anal. Appl.* **42** (1973), 684–760.

Sheffer, I. M.
[1] Some properties of polynomial sets of type zero, *Duke Math. J.* **5** (1939), 590–622.
[2] Note on Appell polynomials, *Bull. Amer. Math. Soc.* **51** (1945), 739–744.

Shohat, J.
[1] The relation of the classical orthogonal polynomials to the polynomials of Appell, *Amer. J. Math.* **58** (1936), 453–464.

Slater, L. J.
[1] "Generalized Hypergeometric Functions," Cambridge Univ. Press, London and New York, 1966.

Steffensen, J. F.
[1] The poweroid, an extension of the mathematical notion of power, *Acta Math.* **73** (1941), 333–336.
[2] "Interpolation," Chelsea, Bronx, New York, 1950.

Szegö, G.
[1] "Orthogonal Polynomials," Amer. Math. Soc., Providence, Rhode Island, 1978.

Ward, M.
[1] A certain class of polynomials, *Ann. of Math.* **31** (1930), 43–51.
[2] A calculus of sequences, *Amer. J. Math.* **58** (1936), 255–266.

Whittaker, E. T., and Watson, G. N.
[1] "A Course of Modern Analysis," Cambridge Univ. Press, London and New York, 1978.

Yang, K.-W.
[1] Integration in the umbral calculus, *J. Math. Anal. Appl.* **74** (1980), 200–211.

Zeilberger, D.
[1] Some comments on Rota's umbral calculus, *J. Math. Anal. Appl.* **74** (1980), 456–463.

INDEX